RATIONAL CHOICE
AND POLITICAL
POWER

Keith Dowding

BRISTOL
UNIVERSITY
PRESS

This edition with new chapters published in 2019 by

Bristol University Press
University of Bristol
1-9 Old Park Hill
Bristol
BS2 8BB
UK
t: +44 (0)117 954 5940
www.bristoluniversitypress.co.uk

North America office:
Policy Press
c/o The University of Chicago Press
1427 East 60th Street
Chicago, IL 60637, USA
t: +1 773 702 7700
f: +1 773-702-9756
e: sales@press.uchicago.edu
www.press.uchicago.edu

The original hardback edition is published by Edward Elgar who retain their rights.

British Library Cataloguing in Publication Data
A catalogue record for this book is available from the British Library

Library of Congress Cataloging-in-Publication Data
A catalog record for this book has been requested

978-1-5292-0633-3 paperback
978-1-5292-0634-0 ePdf
978-1-5292-0635-7 ePub
978-1-5292-0636-4 Mobi

Cover design by blu inc, Bristol
Front cover image: stocksy
Printed and bound in Great Britain by CMP, Poole
Bristol University Press uses environmentally responsible print partners

For Jeffrey William and Sheila Leanora

Contents

List of Figures

Preface

I would like to thank a number of people who have directly contributed to this book by making me add or change at least one and in some cases many sentences. For comments on the original book, now Chapters 1–8 of this edition, that includes Patrick Dunleavy, Barbara Goodwin, Steve Harrison, Richard Kimber, Alan Manning, Brendan O'Leary, Christopher Pollitt and Martin Smith. I also learned much from the members of the 'London Rational Choice Group' at our irregular meetings. These were the most fun and educational meetings I ever attended. They were usually held at Brian Barry's flat in Bloomsbury, and lasted two or three hours, fuelled by the cheese that Brian and Anni provided, and the wine the rest of us brought. Twice there I delivered some of the ideas in this book and the ensuing discussions helped enormously, as did discussions of a paper on this topic delivered at various universities in the United States in early 1987. Members of the Brunel Economics department were helpful in discussing some of the bargaining aspects of political power. I also gave a paper at the Urban Political Studies Group. The lively discussion that followed may not have made me change my mind on any issue, but it gave me fair warning of the problems that I later faced over the oxymoronic – some might say 'moronic' – concept of 'systematic luck'.

Brian Barry, who later became a dear friend, was encouraging from the time of the first proposal to Edward Elgar. Talking to Brian was always a treat and if sometimes I took months to understand his occasionally cryptic comments, their revelatory nature was no less rewarding when I eventually caught up with him. Over the years Brian and I would discuss all sorts of things together. My wife, Anne, and Brian's wife, Anni, would tell of hearing from another room their respective husbands shouting furiously down the phone, only to see them all smiles after the call; Brian and Keith had been having another of their little academic debates.

In more recent years, following Brian's death, my former PhD student, Will Bosworth, has taken on this role. Our discussions these days mostly take place on Messenger, and they can be almost as furious. In the Postscript I give special thanks to Will for his comments on the new Chapters 9 and 10. Not only he did he correct some errors in reasoning, he also reminded me of my own position on certain issues and why I hold them. I would like to thank Anne Gelling for sub-editing the entire book, both original and new chapters. She not only makes me rewrite various passages, but also sometimes causes me to reconsider what I was trying to say. I have discovered that ideas are best expressed in the sentences which contain them rather than ones which I sometimes write in their stead. If there are ideas contained here which you find puzzling then you must blame me, for there are some that Anne, despite my rewriting, still finds puzzling too.

Keith Dowding, Canberra

Introduction to the New Edition

This book was first published in 1991. It appeared in hardback and its initial impact was limited, as I recount in Chapter 9. However, over time it has had a greater impact on the literature, particularly bringing attention to the problem of collective action to those interested in power and the acquiescence of the weak to the dominant. I am pleased that Stephen Wenham, at Bristol University Press, was keen to see this book republished in paperback and e-form. My original publisher, Edward Elgar, retains the hardback rights but generously gave me the paperback and e-book rights. Edward Elgar had come into my office in my first year as a lecturer at Brunel University and asked if I had any book plans. Almost off the top of my head, I'd suggested a book using the tools of rational choice theory to examine political power. Edward was immediately interested, and I produced a book proposal that was enthusiastically endorsed by his reviewer, Brian Barry.

The idea of the book came out of my Oxford DPhil thesis, on Mancur Olson and the logic of collective action. In fact, it came out of one paragraph in that book, an aside, where I had suggested that the collective action problem could explain why people often seemed powerless. We did not need to posit contrary forces, ideology or behind-the-scene activities; people could be powerless all of their own. I wrote this up as a paper, 'Collective Action and Political Power', and gave it at various universities in the USA in early 1987. At that time, the Thatcher higher education cuts were in full swing, and there were few new academic jobs in the UK. I wrote to anyone in the States with whom I had had some contact and who could arrange for me to give a seminar, and contribute to my expenses, hoping to bring myself to the attention of academics. I touted my wares at Yale, Penn State, University of New Orleans and Tulane University, University of California Davis, Chicago University and Indiana University, Bloomington, returning with two potential one-year positions. In fact, however, I was then appointed to the Brunel University position.

Whilst it was not the only paper I gave in the US, 'Collective Action and Political Power' was the one I usually delivered. It was well received in Chicago at the seminar run by Jon Elster, Russell Hardin and Howard Margolis. Jack Nagel at Penn State was really encouraging. At Bloomington I enjoyed wonderful hospitality from Vincent Ostrom, and also met Jeff Isaac. His book on realism and power was more influential on my thinking than he imagined as we argued in his office and over coffee. At Yale what I remember most was giving the paper in a small building in New Haven on a snowy day. Just as I was about to start, a tall man entered, took off his coat and listened attentively. I basically gave the paper to that one man, since I recognized him as Robert Dahl. I was somewhat surprised by his reactions, as he seemed to nod or frown at the wrong times. At the end, he left quickly so I could not speak to him, but I asked the Chair who he was, hoping I could be introduced later. The Chair said, 'I have no idea. I think he was a homeless bum who had just come in to get out of the cold.' I later met the real Robert Dahl, who loved the story.

In many ways, I was motivated to write the book by my response to Steven Lukes's highly popular and influential *Power: A Radical View* first published in 1974. In that book he argues that there are three faces or dimensions of power, and we need to understand them all in order to understand the power structure. I believed that the collective action problem could actually do all, or at least most, of the work of the three dimensions. I also did not like Lukes's – or any other – account that seems to make power so ubiquitous and mysterious. Whilst I am a political philosopher, I am also an empirical political scientist, and I want to be able to analyse and measure power in society. I thought Lukes's and other (even more) 'radical' accounts make power unmeasurable. I reject this for both academic and normative reasons.

First, it seems to me that science is about analysing and explaining. In order to do either, one's concepts must be measurable, and we must be able to model the mechanisms and processes that operate in the world in order to make scientific predictions (Dowding and Miller, forthcoming) and thus explain the world (Dowding, 2016a). Any account of power that makes it unmeasurable is unsatisfactory to me. Second, if we do want to change the world – the point of radical or critical theory – we have to be able to turn our scientific predictions into ones that can guide us as to how we can intervene to make the world a better place. The problem with critical theory, discourse analysis, postmodernism and other self-proclaimed radical approaches is that they do not give clear policy advice, because too often they do not identify specific structural or causal forces of inequality of power. 'Collective Action and Political Power' argued that everything Lukes suggested had to go into the third, unmeasurable, face of power, could in fact be explained by the collective action problem.

That gives us a way into explaining the power structure and why so many people lack power.

Whilst I now think we do need more than the collective action problem to explain some biases in society (Chapter 7 does consider this a little), such biases do not, in my account, constitute the exercise of power. Of course, agents can utilize such bias as a power resource (under my category of information or under unconditional incentives to affect incentives), but it is not itself power until it is wielded. These biases might well enter into our very identities and thought processes, but analytically we should keep them separate from our analysis of power. Social and political power is best seen as a resource that agents can use. Whilst they might use it consciously or non-consciously, that is, the by-product of their actions is still the result of the exercise of power. Nevertheless, some biases in the way we think should not be bundled up with the concept of power as such. I do not think that my claim is simply a verbal dispute over the extension of the concept of power. I think it marks some very different considerations in our analysis of the structure of society. Those considerations entail that we need to intervene in different sorts of ways to bring structural change rather than in merely changing the resources that enable people to exercise power. In my view, distinctions in social concepts are justified to the extent that they do some work. And I think the distinction I mark here potentially does some very important work. The 'radical' approaches tend to miss out on this distinction. I say a little more about this in other places (Dowding, 2006, 2016b, 2008) and discuss it more fully in the Postscript.

In Chapter 8 of the book I promote what Chalmers (2011) dubs the 'subscript gambit', to suggest that we can distinguish different accounts of political concepts so that we do not talk past one another. The chapter attacks the radical relativism that many pull out of the claims of the essential contestability of political concepts. I still maintain that the subscript gambit can help clear up misunderstanding, though I now think that there are even deeper problems with some of the concepts that Gallie's essential contestability thesis implies (Dowding and Bosworth, 2018). Terms can be ambiguous and they can also be vague. We can use the subscript gambit and coding decisions to help deal with both of these problems. However, for many vague and complex political terms, such as freedom or liberty, the subscript gambit and coding decisions bring out the fact that some political terms in our natural usage are in fact incoherent, in the sense that we have contradictory intuitions about what they refer to. The only response to this incoherence is banish the term from scientific discourse replacing it with different terms that map better to our intuitions (Bosworth 2016). We might be able to weight those new values as to how we want to proceed. Our original terms, such as liberty, can then only be used as vague placeholders in natural discourse (Dowding and Bosworth 2018). As it happens, I do not think power is incoherent in that sense that I think liberty is; its different uses are due to ambiguity and not to vagueness. Or, at least, I think by concentrating upon the relative power of agents in terms of their resources we have a clear way in which we can use the term to refer precisely. I argue this more in Chapter 10 in the Postscript.

So, to some extent, at the heart of the book is a critique – a critique of non-empirical and relativist accounts of power. But from that critique I draw out a positive account of what power is. I don't really define power in the book. Rather, I suggest a way in which we can go about analysing or measuring it. Indeed, I am not sure that providing definitions is ever strictly useful. The meanings of terms are better revealed through analysing them, than providing a semantic definition. The account I give allows us to measure the power of agents in terms of the resources which they command. In the book I characterize how to see how resources lead to agential power at a high level of abstraction, based upon the game-theoretic and conceptual work of John C. Harsanyi. However, when it comes to empirical work, we can more feasibly measure the material resources of people: the money at their disposal, the authority they command, the coalitions they can form around themselves. That does not make measuring their power easy, but it does make it possible.

However, as critics are quick to point out, having resources is not the same as having power. People can choose not to wield their resources, they can wield them effectively or ineffectively, and their resources are always relative to others, and how others in turn deploy their resources. However, saying that we can measure the power of agents, and analyse the power structure in terms of the resources which agents command, does not commit what Peter Morriss (2002) calls the vehicle fallacy (Dowding, 2017, ch. 4). This says that power must not be confused with its vehicle. True, but we can still measure power by agents' resources, understanding that these can be used or not, and wielded well or badly. That leads us to a strategic or game-theoretic understanding of power.

Modelling actual processes in this manner requires dynamic game theory, and careful process-tracing. This book does not do that; rather, it analyses the concept of power in a comparative statics manner. It is a structural account. In the opening chapter I look at different ways of examining power, as agency and through structure. I was frustrated at the time by the way that social scientists seemed to think that we either looked at individual agents (methodological individualism) or at structure (holism). It seemed, and still seems, obvious to me that we need both. I thought this was a dead debate in 1991. Apparently not, since people still debate it. List and Spiekermann (2013) essentially spell out the same argument as I do, so it clearly still needs to be made.

The most controversial aspect of my argument turned out to be the account of systematic luck (Barry, 2002, 2003; Lukes and Haglund, 2005; Hindmoor and McGeechan, 2013; Dowding, 1999; 2003). I will say no more about the ensuing debate here, and touch upon it only slightly in the Postscript. Once again, however, the point of the distinction between systematic luck and the systematic power of agents concerns how we should as a society overcome inequalities. If one group is systematically lucky, then we cannot look to disempower them, since they do have not power but luck. Instead we must change the process that makes others systematically unlucky in relation to them. Groups can also be systematically powerful and lucky; again, disempowering them will not necessarily lead to more equal outcomes if their systematic luck remains untouched. So, despite being an analytically philosophical account of the power structure, the whole point of such an analysis of power, normatively speaking, is to provide the weapons to change society. In that sense the book is as radical and critical as any other account of power, even if it does not make that claim explicit.

PART I
The Original Text

1

Introduction

1.1 Introduction

'Power' is an elusive concept. Political power especially so. But, whereas natural scientists do not unduly worry themselves in defining the concept, social scientists do. Natural scientists measure the power of objects in terms of their capacities to do certain sorts of things. This, I shall suggest, is the way in which social scientists should deal with political power. But social scientists are at a disadvantage compared to their natural brethren. The subjects of their study are actors, and that means the subject herself chooses the way in which to behave and is not *merely* caused so to behave. This makes the quest for underlying causes harder, for there are fewer regularities from which to begin the search and those which do exist are harder to interpret. It also makes the study of political power much harder than the study of physical power. Sentient beings have some choice over the manner in which they behave. This makes measuring power more difficult. How *much* choice individuals and groups have is a major point of contention between competing conceptions of political power, as we shall see.

There are many competing definitions of political power, but none is authoritative. There is little point in giving a full-blown review of them all because providing a definition is not enough. We also require a method of demonstrating power in society. We want to know how much power different people have and why they have the amount they do. We want to know the outlines of the power structure. I do not provide a systematic review of the definitions of political power; most of the positions will emerge as I develop my own argument. This argument tries to explicate the important issues. The main positions have developed around a key dispute: behavioural methods of political analysis. On one side behaviouralists believe that they have demonstrated how individuals and groups may affect the policy outputs of government. On the other side 'radical' opponents believe that they have demonstrated that individuals and groups cannot affect the important issues in any significant way. Power is exercised long before individuals try to influence policy outputs. By the end of the 1970s the debate had become very sterile, with both sides apparently believing that they had won. The contention of this book is that if we use the tools of rational choice analysis we can see more clearly what the fighting is all about. Once we have revived this debate we can then ask more coherent questions to try to resolve issues of power in society. I thus give an analytic account of power before turning to a more empirical analysis which considers some of the classic texts of the main power debate in the light of my analytic analysis. This demonstrates how rational choice methods help us to see the issues more clearly. Then the book turns to more recent discussions of power in society.

Chapter 1 is largely concerned with providing a framework into which competing theories may be placed. My own two-part definition does not appear until Chapter 4, which may seem a little late, but we have to cover a lot of ground first. We need to discuss some of the problems involved in conceptualizing power, to look at certain questions concerning methodology and, importantly, to discuss the rational choice approach in general. The latter will be done in the first part of Chapter 2, which should be carefully compared to what I have to say about structural versus individualist accounts in the next section. One of the advantages of rational choice methods is that, whilst they are individualist in the sense that they assume only individuals act, they provide a good entry point into studying the structures which lead people to behave as they do.

As I will not be actually defining political power until Chapter 4, I will give a short preview of some of the conclusions that will eventually be reached. I will identify two sorts of power: (1) 'outcome power', which refers to the ability of an actor to bring about some outcome, and (2) 'social power', which refers to the ability of actors deliberately to change the incentive structures of others. Both sorts involve people working together in coalitions to achieve these ends. This means we have to distinguish group power from individual power. At times it is worth while to think of a group acting as a whole; at others we need to break the analysis down into the actions of individual members of the group. I will identify 'luck' as something distinct from power which leads some to benefit without having to act. Thus the important distinction between 'who has power' and 'who benefits' is demonstrated. It is possible to have luck without power, power without luck, to be lucky and powerful or unlucky and lack power. I will also suggest that luck may be systematic in the sense that some groups have more luck than others because of the sort of society in which they live. This has led many analysts to think of 'systematic luck' as a type of power. I will argue that this is an important mistake. A mistake because belonging to a group or class of people who benefit from the way society is simply does not mean that that group or class is powerful or makes society the way it is. It is important, for understanding this distinction between luck and power affects the manner in which we try to demonstrate power in society. One of the reasons for dismissing behaviouralism is that it is unable to demonstrate the power of some groups which benefit from the way society is structured. Non-behaviouralists have thus argued that behaviouralism must be an inappropriate tool for discovering who has power. However, if what these groups have is not power, but something else, then the inference that behaviouralism must be false fails. Though, as we shall see, behaviouralism as it has been conducted also makes unnecessary assumptions about power. Ideology is different from power, though it may aid both outcome and social power. In that sense it is a power resource. However, I suspect that none of this makes much sense to the reader until it has been built up into a coherent argument.

The rest of this chapter is concerned with laying some groundwork. It will distinguish between various approaches to power. In fact, most of the differences in the way power is defined are *methodological*. But these methodological divisions are not clear-cut and occur at three different levels. Section 1.3 looks at these methodological differences. The next section deals with definitional or conceptual distinctions.

1.2 Definitional divisions

There are two major divisions over the definition of power (though there are also many definitions within each division). The first is between causal definitions and dispositional definitions. The second is whether power is held by individuals or structures. Those who believe that power is structurally held tend to hold a dispositional definition as well. Those who hold that individuals are the wielders of power may hold either a causal or a dispositional account.

Causal versus dispositional accounts

I have already given a commitment to a dispositional account of power in my opening paragraph, where I stated that the power of an object is its capacity to do certain sorts of things. Having the capacity to do something does not mean that the object ever actually does it: it can, under certain conditions, do so. The capacity of an object is a dispositional property of that object. Dispositional properties are strange creatures. Rescher (1975, 132), mixing metaphors somewhat, describes them as 'amphibious' because they 'have one foot in the realm of the actual, another in that of the possible'. The realm of the actual includes objects which exist and events which happen. The realm of the possible includes objects which do not exist and events which do not happen. Dispositional properties such as 'fragility' are actual in the sense that they are present in existing objects, but are possible in that they take their reference from events which may not happen. A fragile cup is one which may easily be broken, not one that is actually breaking. A soluble object is one which may dissolve in water, not one that is actually dissolving. Dispositional properties are thus always theoretical in character, and when we impute a dispositional property to an object we are imputing something which may never be proven. It is for this reason that some analysts prefer causal definitions of power (Polsby, 1980, 60–68).

Robert Dahl (1961a) chose to utilize a causal definition of power, despite having originally defined the term dispositionally (Dahl, 1969b, first published 1957). His reasons were clear. He found it impossible to discover the capacities of individuals in his chosen field study. So he substituted a causal definition for the dispositional one:

> for the assertion 'C has power over R', one can substitute the assertion, 'C's behaviour causes R's behaviour'. (Dahl, 1968, 410)

Similar definitions may be found in Simon (1969) and Nagel (1975). The simple problem with the causal definition is that it is false. 'C has power over R' just does not mean that 'C's behaviour causes R's behaviour' any more than 'this cup is fragile' means 'this cup is breaking'. The error may be called the 'exercise fallacy' (Morriss, 1987, 15–18). The felt need to commit it occurs because of the ancient creed of *operationalism* (Bridgman, 1927) which holds that all physical entities, processes and properties must be defined in terms of the operations by which they are apprehended. In political science this creed led to *behaviouralism*, which I discuss in Chapter 2. Operationalism was a method of logical positivism in which the importance of theory prior to measurement was not fully understood. In Chapters 2 and 3 I shall argue that Dahl found it impossible to discover individuals' power capacities because he did not have an adequate theory of action. Without theories prior to analysis we cannot hope to discover dispositional properties which are theoretical in character.

When we say that an actor has the capacity to do something we are stating that if certain conditions obtain she can do that thing. Whether she will do it is dependent upon other factors: whether or not she wants to, for one. When we look at the capacities of actors in order to measure their power, we are forced to look at those of their properties upon which their capacities are based. We look at the *resources* they have. But looking at these resources is not enough to measure their capacities, for those resources will only allow actors to bring about outcomes under certain conditions. Thus an actor's power is difficult to quantify. This problem of resource-based accounts of power will be returned to in this study. However, the difficulty of quantifying power should neither lead us to ignore it, nor to alter conceptualization or definitions because of it.

Structural versus individualist accounts

In order to understand the debate between structural and individualist accounts of power we must understand what is meant by social structures. Here we find immediate difficulties for there is much wider disagreement over conceptions of structure than there is over the nature of dispositions, causes or actors. I will produce what I take to be the clearest account of structure, though many structuralists will believe it to be far too simple. I do not attempt in this book to give a full-blown critique of complex and in my view confused accounts of structuralism, though what I do say should give a hint as to what such a full-blown critique would look like.

The structure of anything is the relationship its constituent parts bear to one another. The structure does not exist above or beyond those constituent parts – if there were no parts there would be no structure – but the structure is logically independent of the parts. It can be discussed separately. Cohen (1978, 36) writes:

> One may know what the structure of an argument is without knowing what its statements are, and one may know what the structure of a bridge is whilst being ignorant of the character of the parts. One may, moreover, remove the original statements and replace them with others without changing the argument's structure, and the same applies to the structure of the bridge, though the second operation requires great caution.

The constituent parts of the social structure include individuals. However, it is possible to discuss the nature of the social structure of a given community without giving a proper name to any of the constituent individuals. Similarly, it is quite possible to discuss the power structure without actually giving the proper name of the power holders. We may denote the power holders by their place in the social structure alone.

As power is a dispositional property of an object, it is also structural in the sense that its nature derives from the relationship between certain properties of that object and properties of the external environment. However, it is a property of the object and not of the relationship between the object and the external world. 'Power' is both a property of objects and a relation between them. We usually refer to 'political power' as a property of individuals or organizations (or sometimes unorganized groups of people) which act. The actions whose consequences demonstrate these powers are often intentional but may be unintentional. Power can be both a property of objects and a relation between them if it is a property-relating disposition (*pace* Hindess, 1982, 505). If A can make B do something, then A has power over B with respect to the something she can make B do. That power is a property of A, and is a power property because of the sort of relation it is. Now it may be true that A only has that power over B because of other relations they bear to each other and has nothing to do with A denoted in any other way. These other relations define the scope of A's power. Thus Dowding has the power to make Smith write her essay on political power because he is a university lecturer and Smith his student and not for any other reasons. If Smith were the lecturer and Dowding the student the power relation would be exactly the same, only now with Smith the tyrant and Dowding the victim. Nevertheless, the power that A has is a property of A and not of the relationship, in the same way that the fragility of the cup is a property of the cup and not a property of its being dropped. The power structure is a description of the power properties of all the individuals who happen to exist. We may discover why these individuals have the powers they do by examining the other properties they have, including other property-relating dispositions.

Thus we may explain the power structure by examining the relations between individuals and in our *models* of society we denote actors by their relationships to other objects and actors and not by their *proper names*. Individuals act as they do, not only because of their wants and desires but also because of the opportunities they perceive to be open to them. This perception is directly related to their position in the social structure. I think that this is what leads some analysts to believe that power is a property of social structure. Such analysts tend to concentrate upon the closure of opportunities that each individual's choice situations produce. That some individuals have fewer opportunities than others is said to be due to the power of the social structures (or 'system') to close down certain opportunities and thus restrict choice. But in defining the lack of social power of the individual the analyst is defining the lack of social power of the individual and not the power of the individual's relation to others (which is the structure). Individuals' relations to objects and other individuals cannot have power (Lukes, 1977, 9).

Other analysts write of social structures 'enabling' people to do things rather than (or as well as) social structures closing down opportunities for them (Giddens, 1984). Thus, an individual may be said to have been 'enabled' by his public-school, Oxbridge education to attain the higher reaches of the British Civil Service. Belonging to the 'right' social class, having the 'right' social manner enabled that individual rather than his state-school and redbrick-educated but equally formally qualified counterpart. The distinction between social structure's 'enabling' and its 'closing down' opportunities is too fine for my understanding. If there is some opportunity, some job or role to be fulfilled and A is enabled to take that opportunity then it must be at the expense of some other person B. Rather, what we mean by social structures closing down opportunities for some or enabling for others is that some people have an advantage over other people in doing or becoming those things for which they are in competition. There is no doubt that this occurs, but it does not follow that the social structures themselves have power to bring these things about, just that the way in which society is structured gives competitive advantage to some people over others. Indeed, outside of a pure lottery, it is difficult to see how society could be structured so that some people do not have an advantage over others for those things for which they are in competition. Individuals are bound to have different sets of properties depending upon the relationships they bear to one another. My subject is not, however, the justness of different sets of advantages, but the identification of these relations. The only sense in which enablement does go beyond constraint is where individuals gain properties which give them the ability to do things which would not otherwise be done at all. But it is the property which gives them this power and not the structure as such; the latter is merely the description of the relationship between different people who may be denoted by the properties they have.[1]

I think it is wrong to ascribe power to social structures. It is a mistake to think that because we are mapping the structure *of* power, that structures *have* power. Describing the distribution of power in society by the relations between people does not mean that the relations between those people are themselves powerful. The theory that structures have power may be dismissed by two arguments: first, such ascriptions are redundant; and second, they are misleading. The first may be called the *redundancy* argument, the second the *conceptual* argument.

Redundancy argument

It is claimed that many of the freeways designed by Robert Moses for New York embody racial and class bias (Winner, 1980; Ward, 1987). Low-level bridges prevent buses from using certain routes: those sections of the population which do not own cars are therefore effectively prevented from visiting certain areas of the city. Do these physical structures have the power to stop certain individuals visiting parts of New York? Not alone. In order to show why these individuals cannot visit some districts we also have to explain that they do not own cars. Leaving that aside, what we have here is a lack of power on the part of individuals, not a structural power of the freeways. If I cannot go south because there is a wall in the way, then we may wish to say that the wall has the power to stop me going south. But we do not *need* to. All we need to say is that the wall stops me going south, which is a less long-winded way of saying the same thing (conceptually the sentences are equivalent). The use of the word 'power' here is redundant. Moreover, it is misleading.

Conceptual argument

The use of the word 'power' here is very different from when we talk of the power of actors, and different in a way which makes the use of this abstract noun misleading (conceptually the usage is different). When we talk of the political power of individuals we are talking of a power that they can (and may) choose not to wield. If A has the power over B to the extent that she can make B do *x*, then she can make B do *x* but she may choose not to bother. Structures do not have this ability to choose not to wield power. If we are to use the phrase 'structural power' we need to carefully demarcate it from ordinary uses of 'political power'. Now note that, whilst individuals can choose not to wield power, their own choice situation may make that non-wielding option unappealing. The fact that the structure of individual choice situations may make outcomes predictable with a high probability does not show that structures have power. It merely demonstrates that the power of individuals is in part determined (or rather structurally suggested) by their positions in the social structure. The conceptual distinction still holds.

In fact, the Moses freeway example is not a very good case for structural power. Moses helped set up the Triborough Bridge Authority to produce a controlled number of toll roads and bridges in a high-volume traffic area. This created a set of captive consumers buying a private good, without taking into account the negative externalities of pollution, environmental damage, urban sprawl and over half a million people displaced by highways. By creating a quasi-private authority, the Moses empire was set aside from public scrutiny. How far Moses was aware of the system he was setting up is moot; he may even have deliberately created the inequalities his roads produced (Caro, 1974). This would enable us to say that it was Moses' power which caused the inequality. Certainly Moses is a prime example for many elite theorists (Molotch, 1976; Stone, 1987). However, even if the inequalities were an unintended consequence or by-product of Moses' actions (which I take to be the point of the example), the structure does not have power, for the reasons given. It is simply that the structures reduce the powers of some groups to take certain actions. A reduction in one person's power does not entail an increase in the power of someone or something. (To think so is to commit 'the blame fallacy': see Chapter 5.)

1.3 Methodological divisions

Alongside the definitional divisions are divisions over the best way to study power in society. These divisions appear at various levels and we will begin with the grandest one.

Methodological individualism versus holism

This division mirrors the division we saw between structural and individualist definitions of power. Individualist accounts suggest we must look at individuals in order to understand the power structure. Holists suggest that only to study individuals is to miss out most of the important features. Both sides have valid points, and I will suggest that we must be individualists but that we must not be methodologically so.

There are many methodological individualists and not all defend the thesis in quite the same way; but I take the main idea of methodological individualism to be that all explanation of social phenomena must be given in terms that are reducible to propositions which contain reference to individuals alone and not social wholes.[2] The explanation of a political outcome by reference to the actions of the Prime Minister, her party officials and the international banking system would be reducible to propositions containing reference to individuals alone, since the actions of each of these institutions are caused by the actions of individuals within them. Careful individualists admit that the reasons why individuals act may depend upon their structural properties. The Governor of the Bank of England makes decisions as Governor, rather than as Robin Leigh-Pemberton. In doing so he will take into account the interests which he is supposed to defend. These are the interests of the Bank, of the government and of Britain, rather than his own individual interests, which may be at variance with some of these. But individualists claim an advantage over some holistic accounts, for their methods allow that the governor may be swayed by personal interests which may go against the institutional ones. For example, Niskanen in his models of bureaucracy assumes that bureaucrats work in the interests of the bureau (and thereby their own personal interests) rather than in those interests which the bureau is supposed to further (Niskanen, 1971; see Dunleavy, 1985, 1989 for a more sophisticated rational choice model of bureaucracy). Careful individualists will also admit that institutions exist in a primary sense, and embark on a causal account of outcomes by entering into the thought processes of individuals. Thus an individual may become a communist revolutionary because of her belief in the true nature of proletarian interests.

Careful individualists have thus come a long way towards their holistic rivals from early definitions of methodological individualism (von Mises, 1949) but not far enough to satisfy all holists. Most holists now agree that only individuals act (the basic individualist thesis) but suggest that individual *reasons for action* cannot be the prime explanatory variable.[3] If the individualist admits that social wholes or institutions enter into explanations through the beliefs of individuals, why are the beliefs rather than the institutions given explanatory primacy (Nozick, 1977)? The reason, I think, is that the beliefs are what motivate action – the vehicle of causation – and the beliefs may not be a true representation of the institution. The existence of the institution may help to cause the belief which leads to action, but the content of the belief about the institution may also be partly caused by individual desires (Williams, 1972; Elster, 1983a).

Modern holists make a more fundamental criticism which attacks the reductionist element of methodological individualism. Methodological individualists want to *replace* macro-level holistic explanation of phenomena with micro-level individualist explanation. Holists suggest that some good macro-level explanations cannot just be replaced with micro-level ones. They explain this by introducing the *type/token* distinction (Levine et al., 1987). A token is a specific example of a general class. A type is the general class which is made up of many token examples. Any given token may belong to many different classes, and any type may have many different token examples. 'Thus a particular strike – a token event – can be subsumed under a variety of possible "types": *strikes, class struggle, social conflicts*, etc. Similarly, being rich is a type of which Rockefeller is one token' (Levine et al., 1987, 76). The holists then suggest that not all type explanations can be reduced to token explanations. Whilst social science wants to explain why certain token events occur, it also wants to explain the nature of certain types by which we explain the occurrence of token events. Thus the 'fact' of class struggle enters into an explanation of the General Strike of 1926, or the dominant ideology enters into an explanation of the quiescence of the working class in modern Britain. These 'facts' cannot be further reduced.

Again, however, the individualist has a reply. He admits that these social types are used in explanation, and may also admit that a part of social science is explaining these types and is indeed prior to token explanation. The first may be called 'descriptive theory' and it is logically prior to 'explanatory theory' (Stinchcombe, 1968, 55; Nagel, 1975). But the individualist claims that individuals must be used in explaining the nature of these types, and certainly used in explaining their formation in actual society. The holist replies by introducing the concept of *supervenience.*

Water is H_2O. Explanations using the term 'water' may be reduced to ones using only hydrogen, oxygen and the laws governing their bonding. But holists claim that not all relations of dependence are like that. Some types are sustained by very different tokens. Different surface properties in different conditions may lead me to see the same colour (say 'red'). Different brain states sustain the same belief (say 'I am cold'). Different properties of an organism and features of its environment explain why this organism has the evolutionary fitness it does (Sober, 1984, ch. 1).[4] Similarly, different beliefs, resources and inter-relationships may realize the same social type. The Labour Party may get 34 per cent of the vote at two different elections, but that same result is produced by different people with different sets of reasons for voting Labour. In each case, the colour, the belief, the fitness and the Labour vote is said to supervene upon its constituent parts. Because the *same* supervenient state may depend upon *different* subvenient constituents, the constituents are not important in any particular explanation in which the supervenient state is used. Thus the fact that the bull charged at me may be explained by my red shirt, and there *need* be no reduction as to why the shirt appeared red to the bull. Frogs and giraffes both survive today (whereas brontosauruses and moas died out) because they are evolutionarily fit in the environment through which they have lived. In the first case we need make no reference to the subvenient constituents in our explanation of a token event. In the second case we need make no reference to the subvenient constituents in order to explain the type outcome.

There is some merit in this argument, but it need not unduly worry the reductionist individualist. It is true that we use colours and not their atomic bases in many explanations. My beliefs explain my action and not my brain states. But we also use water in many explanations rather than reducing it to its constituent parts. But in each case we *could* so reduce, and that is all the reductionist claims. All macro-level explanations *can* be reduced to micro-level ones.

The key to why we use macro-level descriptions rather than micro-level, however, is that the macro level is all that we require. What matters here is the interest of the questioner. Putnam points out that we could produce a geometrical explanation and an atomic explanation of why a square peg will not go into a round hole. Both are equally correct answers. Why we give one rather than the other depends upon the nature or level of the question being asked (Putnam, 1978, 41–5). Macro-level explanations are often all we require. The question as to why the Labour Party is the party of opposition and not the government may be answered by saying it only got 34 per cent of the vote. But often we ask deeper questions. When we ask, 'Why is the Labour Party not the party of government?' we may be asking why it only received 34 per cent of the vote. Here we do need to probe into why those people who voted for it did so, and why those who did not voted the way they did. But again, the answer to those deeper questions will also involve sentences using types. Type explanation here may also be sufficient, but a further question may be asked about these type explanations.

So who are right, the reductionists or the holists? I am going to take up what may be a rather uncomfortable position in the middle. I will assume that all type explanations may be further reducible, and that we may wish thus to reduce them. But often we will not want to go further. And this is not just for the contingent reason that we have finite minds. If I ask God 'Why am I here?' I do not want Him to give me the entire history of the universe in all its detail up to the moment I asked the question. And I would not want that answer even if I were an infinite being with plenty of time to spare. Much of such a reply would be irrelevant. The problem, however, is that it is not possible to give general criteria of relevance prior to particular questions and particular answers.[5] Hence holistic answers, or at least type answers, may be sufficient for some questions, but I do not think we should rule out the possible relevance of reduction of all types in any given answer. An answer in terms of fitness may satisfy my questions about frogs or giraffes, but I may wish to know why this type of frog is fit even if the answer is different from why this type of giraffe is fit. And I may want to know why this token frog survived when another token frog died. In both cases the evolutionary fitness of frogs may enter into the explanation, but needs to be reduced to the token case.

Thus I am an individualist in the sense that I hold that (a) only individuals act,[6] and (b) explanations of both token and type events may be reduced to sentences using only words which refer only to tokens, or to words which may themselves be reduced to words referring only to tokens. But I am not a *methodological* individualist because often such reduction is otiose. Indeed, often such reduction would not do the job of answering the questions we ask. In this book I am only concerned with explaining types of power and, whilst all my explanations are individualistic, they only denote individuals by their structural properties. Thus I only explain by type. However, explanation by type answers many of the questions we want to ask about the token power structures in our society.

The truth in structuralism

I have suggested that it is a fallacy to regard political power as a property of a structure rather than a property of actors. But this is not to deny truth in some important structuralist claims. Structures exist in the sense that they are descriptions of the relationship between sets of individuals and between sets of individuals and other natural objects. As those relationships change then so do the structures. But an important truth in structural accounts of society is that these structures are relatively enduring. It is not easy to change one's relationship with others, even given the will to do so. Some modern structuralists have taken up a realist ontology in order to explain this endurance (Debnam, 1975; Isaac, 1987). Realists hold that empirical regularities such as 'whenever x then y ...' must be underlain by enduring mechanisms which cause them (Bhaskar, 1979). A realist account of political power explains regularities of power relations in terms of other enduring features. Thus if person A has power to do x then that person has that power because of some of her essential properties. If person A has power over person B (can cause person B to do x) then A has that power because of some of her essential properties. Those properties will be the resources that A may bring to bear on B. Now B has the free will to refuse to do A's bidding – a problem for all institutionalist accounts of power, as we shall see (the 'stubbornness objection' in Chapter 4). But essentially A's resources enable her to bring incentives to bear on B to act as A desires. If the resources of people in society are relatively enduring then so are the probabilities of one set of people bringing incentives in to affect other sets of people. It is because their resources are relatively enduring that the structures may be said to be relatively enduring. Bluntly, if a set of people has greater powers to get things done or have powers over others they are not going to give them up if they can help it. Now note, this does not entail that they constantly fight to keep these powers. If the resources upon which their powers are based are not under threat they may never think about them. But if they do become threatened then they are likely to defend them. In other words, the structures are relatively enduring, for there are few incentives for anyone with power to change them, or even think about them, and it is difficult for those without power to do so. There may be few incentives for the powerless even to try. That does not mean that there are no incentives, however, nor that they will never be changed.

Realism is structuralist because the properties people have which give them power are relational properties; in the example earlier (page 8) Dowding has the power to make Smith write her essay because Dowding is the lecturer and Smith the student. Even money, as material a good as we can have, is a relational property in the sense that the value of money (what we can do with it) depends upon the relative amounts that people have in society. As a resource to be brought to bear in power relations, money-holding is a relational property of actors. Descriptions such as 'lecturer' and 'student' both denote by role and describe the relation between us. Descriptions such as 'rich' and 'poor' describe a relation between people, and may denote a role depending upon other factors. How far the latter descriptions play a role in mapping the power structure will emerge later in this work.

This fact that properties of individuals also denote relations between people seems to confuse some analysts (for example, Hindess, 1982). It is a fairly common denotational error to think that some property, such as power, is *either* a property of an actor *or* a relation between actors. However, *all* unitary properties can be defined in binary terms but not vice versa:

$$(\forall_x) \, (\exists_y) \, P_x \overset{def}{=} R_{xy}$$

which means that, for all x and some y, if x is a P then x is R-related to some y. Thus binary properties are more basic than unary ones. So what? This may seem to be a trivial truth but I think it is an important one. The only way we can denote any object is by its relationship to other objects. The word 'blah' does not denote anything until I describe the thing that I am referring to (or 'naming') by saying 'blah'. I can only do this by describing it in relation to other objects, for example '"Blah" is the thing under the table, the only red object in the room' etc. Giving objects proper names only denotes them when tied to other descriptions. When we come to describe (or 'model') the power structure we describe actors by their relationship to other objects, including other actors. This essentially structural description denotes them as the actors they are. The structural description, because it is basic, is the important one, not the proper name which derives its sense from the structural description. I think it is important to understand this, for the structural description is what explains individual and group power; but that power is still individual and group power and not structural power.

Realism becomes marxist when the underlying mechanisms of power relations are given in terms of *class* relations. The most powerful individuals in society belong to the capitalist class and the powerless to the proletariat. Realist theory has much to commend it and my positionalist account of power relies upon a realist framework. However, unlike Isaac (1987), who gives the best realist account of power, I will suggest that this framework is compatible with a form of behaviouralism. I will not be expounding a marxist account because of difficulties both with its class analysis and with historical materialism. Rather I wish to produce a more subtle group analysis with a more liberal approach to individual interests which retains a contingent view of historical events. Hence my preference for the phrase 'structural suggestion' over 'structural determination'.

2

Rational Choice and
a Theory of Action

2.1 Rational choice

In this chapter I will explain the rational choice method and demonstrate that it is individualistic and yet explains by describing the structures which condition choice. The rational choice or economic approach to the study of politics is now well known in political science. Its proponents proclaim many advantages over 'sociological approaches' (Barry, 1978) though it is not without its difficulties and critics (for example, Hollis and Nell, 1975; Hindess, 1988, 1989). In my view the main difficulty with rational choice is that too many claims are made in its name – usually by its critics. It cannot and is not supposed to provide the final or definitive explanation of political behaviour and political outcomes that is sometimes presumed to be its objective. Rather, rational choice is a way of generating questions about society by modelling social situations. A model of a social situation is just a description of it. A rational choice description of a social situation picks out certain features of that situation and is a good model – and a good description – when those features are the important ones. The important features of a good rational choice model are not the basic assumptions of the rational choice method – which are contained in all rational choice models – but the specified relations between the actors. The model describes the relationships (and hence the individual properties) of all of the actors. A good model should replicate the structure of the situation it describes and tries to explain. The structure of the model together with the rational choice assumptions determine the outcome. How good rational choice is as a method is determined by how good its general assumptions are. How good each model is as an explanation of a particular situation is determined by the structural fit. The assumptions which make the models rational choice ones govern the ways in which individuals make their decisions about how they are going to act. However *it is the structure of the individual choice situations that does most of the explanatory work. It is the set of incentives facing individuals which structurally suggest behaviour to them; by studying those incentives together with assumptions about the way actors make decisions we come to understand why people act as they do.* The worth of a model is measured by its structural correspondence to actual conditions, and not by its behavioural assumptions, which can be varied with greater ease. When models fail because of failures in the rational choice method the method must be supplemented or discarded. When the models fail because of the lack of structural fit then a new model is required. Critics of rational choice theory have been too eager to discard its general assumptions; proponents have perhaps been too reluctant to jettison simplistic models.

The assumptions governing individual decision-making have proved to be the most controversial aspect of rational choice theory. In order to produce non-trivial explanations of human behaviour, rational choice usually operates under the assumption of self-interested utility maximization. It assumes that individuals act as they do in order to maximize their own self-interested utility. But rational choice does not have to operate with such a limited view of human nature. Some models operate under assumptions of altruism with the perceived self-interested utility of others entering into the utility function of actors (for example, Frohlich, 1974; Collard, 1981; Margolis, 1982; Sen, 1982c, 1982d). Handled carefully, such models may produce non-trivial explanation and surprising results. However, I shall in this study generally assume self-interested behaviour, for this assumption is specific enough for my general claims. It is a false assumption but true enough for our purposes. I assume that most people behave egoistically most of the time and altruistically some of the time, whilst some people behave altruistically most of the time and egoistically some of the time. For explanations of mass behaviour the self-interest assumption seems specific enough, though at times particular individuals may have to have their behaviour explained by assuming other types of motivation. Generally speaking the larger the group studied the less unrealistic the assumption of self-interest. Imagine trying to explain the behaviour of a certain individual, call her Susan. In order to explain what she is doing we may have to make many assumptions. We need to understand her beliefs and desires, to perceive what she perceives and the way in which she perceives it. We rationalize her behaviour in much the same way that she rationalizes it herself. This Principle of Humanity (Grandy, 1973) allows us to understand her behaviour as well as she does, possibly better, for we have another perspective from which to criticize the one she has of herself.[1] A stark assumption of self-interested behaviour would not be very useful here and would soon be falsified unless we continually produced schizoid explanation in order to save it. But if we want to explain the activities of a large group of, say, a thousand people we could not begin to explain all of their behaviour in as complex a way as we could for Susan; we are not interested anyway in all of their behaviour, only that which relates to the group. Here, given that most of the people behave egoistically most of the time, egoistic assumptions will get us a long way but where it does not get us far enough then we need some finer-grained assumptions.

A more serious problem for rational choice comes from the assumption of rationality itself, which is often called 'thin rationality' (Elster, 1983a). Essentially, thin rationality means consistency and completeness. Completeness is defined by 'connectedness of binary relations': individuals either prefer one option to another in a choice set or are indifferent between them. Thus for all individuals i and two options x and y:

$$(\forall_i)\ i:\ \{(x > y)\ v\ (y > x)\ v\ (x = y)\} \qquad (2.1)$$

where $\forall_i$ means 'for all i'; $>$ stands for Preference and $=$ for Indifference; and v means 'either/or but not both'. So (2.1) states that, for all individuals i, i prefers x to y, or prefers y to x, or is indifferent between them.

From these binary choices we produce a preference schedule under the assumption of transitivity which gives us consistency. A *preference schedule* or *preference ordering* is transitive under the condition that if one option is related to another in a particular way and that other is related to a third in the same way, then the first option is related to the third in that way. Thus:

$$(\forall_i)\ \{[(x > y)\ \&\ (y > z) \rightarrow (x > z)]\ v$$
$$[(x = y)\ \&\ (y > z) \rightarrow (x > z)]\} \qquad (2.2)$$

For all individuals i, if i prefers x to y, and prefers y to z, then i prefers x to z, or if i is indifferent between x and y and prefers y to z then i prefers x to z. Transitivity is at the heart of rationality. If it does not hold, individuals may rationally 'better themselves to death'. For if individual i orders his options in a preference schedule intransitively: $\{x > y > z > x\}$ then i will be prepared to swap x for z plus some money, z for y plus some money, and swap y for x plus some money and so will end up with what she started with minus all the money she handed over in exchange.[2] One of the problems for rational choice is that individuals do not always seem to have transitive preference orderings. They may seem to prefer x

to y, y to z and z to x (for example, see Tversky and Kahneman, 1981). How this may come about and how it relates to political power will be discussed in Chapters 3 and 7. Another problem for rational choice is that connectedness does not always hold for actual individuals. They might neither prefer one option to another nor be indifferent; they just haven't thought about either option (Sen, 1982a). The problem for rational choice is that it is hard behaviourally to distinguish indifference from unconnectedness. If a person is indifferent between two policy outcomes then she has no incentive to try to ensure one outcome rather than the other. We would not expect her to contribute to any decision furthering one of those outcomes at the expense of the other. However we would not expect her to contribute to any decision concerning one outcome over the other if she had unconnected preferences either. Here, however, the individual might well have strong preferences for one over the other *if she connected the two together*. One of the questions plaguing power debates is how far these two different situations affect an analysis of power. Should we say that someone is powerless because they have unconnected preferences? How do we behaviourally distinguish unconnectedness from indifference? The latter question is approached via the hoary old subject of 'objective interests' in Chapter 3. Some of the issues surrounding the former question are approached in Chapter 7. More complex criteria for consistency may be required to take into account time and probability calculations (Elster, 1983a) but that will be set aside here.

Various broadly 'rational choice' accounts of power have been attempted by others. I think it is true to say that outside fairly closed circles these accounts of power have not had a great influence in political science. I think this is because they do not quite address the right problems. In short what they identify is not power but something else (Barry, 1980). I shall suggest that these accounts are unsatisfactory largely because 'preference' plays either too great a role or the wrong role altogether. My own account will bear more directly on broad debates within political science.

In order to avoid (note, not solve) problems of preference orderings I will be utilizing the concept of individual interest (although individual interest will be explained largely in terms of preference orderings). Interest is a more familiar notion to theorists of power and is not without controversy. But interest and preference will be marshalled together in an examination of individual and group behaviour.

Rational choice models are generally kept quite simple and whilst they can be applied to a variety of social situations they become more useful when other information is added to them. More detail may be added to simple models in order to make them correspond more closely to the situations they describe. This enhances explanation of particular social outcomes. That added detail, however, demonstrates how the players face individual incentives to act in various types of ways. These incentives may change the beliefs and even preferences of the players in ways to be examined later (see Chapter 7).

Modelling may take two general forms. We may model an actual (token) situation (for example, Allison, 1971) or we may model a class of (type) situation (for example, Elster, 1985). If we are trying to explain a particular social outcome by use of a rational choice model we are suggesting that that situation shares some common features with the model. We apply the model and see how closely the assumptions fit. We may modify some assumptions and some aspects of the model to produce a better fit and point out discrepancies between the real-life situation and the model. The discrepancies do not falsify the model for we are making no cognitive claims about it. We are not claiming that the model is true. But the model is useful if it enables us to understand the situation better than another description (for a model is just another description of a situation). Part of the model's usefulness may result from the discrepancies between it and the real-life situation. For those discrepancies may in fact be important in explaining the actual outcome and may be what distinguishes that situation from other similar ones. The discrepancies may be the main causes of the outcome rather than the features of the model itself, which become the 'background conditions' of what Mackie (1974) calls 'an insufficient but non-redundant part of an unnecessary but sufficient (*inus*) condition'. The main features of the model may thus be *inus* conditions of the outcome rather than what is ordinarily called the cause.

Let me give a traditional example. The barn catches fire. What is the cause? In this (token) case the immediate cause of the fire was the match carelessly dropped. But other conditions were necessary for the fire to have started. There was a pile of dry straw, a ready supply of oxygen from the open window and ill-fitting doors (and perhaps lack of firefighting equipment) and so on. The dropped match may be said to be *the* cause of the fire, for without it *this* fire would not have started. It is true that another fire, like it in all respects, might have started at the same moment from the dangerous electrical wiring and in which all the other conditions would have taken the same part. In that case the dangerous electrical wiring may be said to be *the* cause of the fire. It is also true that another match, like the first in all relevant respects (and from the same packet), could have been dropped and started the fire. I suggest, however, that we would not really be interested in which match from the packet had started the fire. We would be interested in why the match (any match) had been dropped, and in the other *inus* conditions in the barn. Rational choice models rarely, if ever, produce causal explanations of actual political outcomes in the traditional sense of cause here, rather they model the *inus* conditions which are equally important (*pace* Elster, 1983b, 25–48).

If we were to model a barn fire we would need to put in all the conditions that were relevant to the actual fire, and those relevant to a possible fire in that barn. These would constitute the structure of the model. If we imagine the fire started by the poor electrical wiring, we might say that the structural conditions were themselves causative (though it would still be an event – the flow of electricity – which caused the wire . to heat up and start the fire). If we imagine the fire started by an outside active agent (someone dropping the match) then the structural conditions are important but are not actually the cause. I think this analogy is a pretty good one for rational choice models of social situations, though there are important differences. The most important is that the structure itself conditions the beliefs and thereby the desires of the actors which are the motive cause of social outcomes. (This may also be true of the barn case. The fact that the barn was poorly lighted, owing to the faulty electrical wiring, may have been the reason why the actor lit the match in the first place.) However this seems to make the structural conditions more, rather than less, important in rational choice explanations. But the barn analogy is how I see rational choice helping us to explain social outcomes.

2.2 A theory of action

Rational choice has not always operated with an explicit theory of action but it has always operated with an implicit one. The implicit theory assumes that individuals will always select the option among those available which they prefer rather than one(s) which they do not prefer. Preferring one alternative to another may also be said to *be* giving it higher utility. The most basic assumption of rational choice is that individuals attempt to maximize their marginal utility. Their actions are thus determined by what they desire and what they believe to be possible, for failing to bring about some highly desired outcome does not maximize marginal utility. Maximizing marginal utility thus includes probability calculations as well, weighing up the expected utility of possible options. Any theory of action must include reference to individual beliefs and desires.

The best and most widely accepted theory of action is that of Donald Davidson (1980) (see also Pettit, 1978; Elster, 1986). For Davidson, an individual's reasons for action may be considered the proximate cause of the action; indeed, those reasons for action denote it as the action it is. If I scratch my nose because it itches, that is one action; if I scratch it to give a signal to an accomplice, that is a very different action. Reasons for action are analysed by Davidson in terms of belief and desire. Desires motivate whilst belief channels the action. So an individual who desires z will do y because of her beliefs x. The three go together in a triangle of explanation and given any two of the triumvirate the third may be predicted and thereby explained. Reality helps to cause actions mediated through beliefs and desires. This is a behaviouralist theory of action, since it is studying the behaviour of individuals that allows us to understand their beliefs (by making assumptions about their desires) or their desires (by making assumptions about their beliefs). We may understand both by making assumptions about different aspects of each (Davidson, 1985).

2.3 Behaviouralism

The way in which I have been explaining rational choice theory suggests that it produces a form of structural explanation in that the main explanatory component is the structure of the models which applied to different situations. Behaviouralism is usually considered to be a rival form of explanation in that it places emphasis upon the actions of individuals to explain social outcomes. However the form of structural explanation I am associating with rational choice uses explicit assumptions about behaviour and in them it is the actions which are the causes of the outcomes and not the structures. The behaviouralist method too relies upon the structural features of situations to explain mass uniformity of behaviour. Behaviouralism is

> a new approach to the *study of political behavior*. Focussed upon *the behavior of individuals* in political situations, this approach calls for examination of the political relationships of [people] – as citizens, administrators, and legislators – by disciplines which can throw light on the problems involved, with the object of *formulating and testing hypotheses*, concerning *uniformities of behavior* in different institutional settings. (Annual Report of the American SSRC 1944–5, quoted in Dahl, 1961b, 764, emphasis original)

The Davidsonian method must be modified when it is applied to mass behaviour rather than to single individuals. More general assumptions about the beliefs and desires of people have to stand for particularistic assumptions. These assumptions lead us to expectations about mass behaviour and are explanatory to the extent that behaviour conforms to our expectations. Further empirical research may then be conducted to test our expectations where they conform and to develop new hypotheses where they differ.

Set out thus, behaviouralism appears an innocuous thesis. It seems difficult to understand how we could have a social science if we did not study human behaviour. However behaviouralism, which is a methodological thesis, must never be confused with *behaviourism*. Behaviourism is the *ontological* doctrine which reduces all mental concepts to publicly observable behaviour (Skinner, 1953).[3] This thesis is long out of fashion (Taylor, 1964). The problem for behaviouralism is that it has had the tendency to reduce to behaviourism without the analysts' apparent awareness. Behaviouralists have not always recognized the need for a complete theory of action in order to understand behaviour and have tended to ignore the need to formulate hypotheses prior to testing them against individuals' actual behaviour. Thus critics of behaviouralism have been led into rejecting the simplistic accounts of human beliefs and desires offered by behaviouralists and to look for explanations outside of individuals. However, armed with an explicit theory of action, this is not necessary. The problem is not with behaviouralism, but with behaviouralism which does not have an adequate theory of action.

Behaviouralism within community power studies in political science is associated with the work of Dahl and his two associates, who conducted an extensive study into the power structure in New Haven between 1957 and 1959 (Dahl, 1961a; Wolfinger, 1960; Polsby, 1980). But their methods have been used in a number of further studies in both Britain and America (Bealey et al., 1965; Bell and Newby, 1971; Newton, 1976). In fact the manner in which the behaviouralists went about their research differed little from how their reputational and positional forebears went about theirs (Ricci, 1971, 128–9). They all interviewed people, read newspapers, consulted archives, established formal and informal contacts with individuals in the community, went to public meetings and so on. The real difference lay in what was selected for investigation.

The behaviouralists chose a causal definition of power already criticized in Chapter 1, justifying this on the grounds of studying 'actual' rather than 'potential' power. They did this in order to avoid the vehicle fallacy. But in doing so they were led to the study of issues already in the news. This has two implications. First, they studied issues that were 'political' or 'public' in the sense that they were already issues which were discussed by politicians in the public domain (Polsby, 1980, 4–5; Dahl, 1963, 254). This is the sense of 'political' that Easton had defined at about the same time (Easton, 1953, 1965, 1966). They may be criticized for having too narrow a picture of the 'political'. Second, they were studying C. Wright Mills's middle level of the power structure (Mills, 1956). This latter is only a criticism of their work if we can justify the hypothesis that there is a level of power not reported by the media.

Dahl and his associates decided that nothing could be assumed about power structures prior to analysis of events (Polsby, 1969; Dahl, 1961a). They thus decided to concentrate upon certain 'key' issues (Dahl, 1969a, 38; Polsby, 1969, 32; 1980, 96). There were four criteria for selecting which issues to study: (a) How many people are affected? (b) What kinds of cost or benefits result? (c) How widely are these distributed? (d) How much is the community's existing pattern of resources affected? (Polsby, 1980, 94–6).

The New Haven researchers recognized that there were leaders in New Haven, but they argued that these people did not constitute an elite in the sense that they did not rule in their own interests and their power was limited. Few wealthy or socially important people were politically important in the key issues studied, and few people were important on more than one of the key issues. Hence resources and power are not cumulative and everyone has access to some of them (Dahl, 1961a, 1 *sic*, 226–8; Polsby, 1980, 119–20): thus no one resource dominates all the others (Dahl, 1961c, 83). Thus they inferred their pluralist conclusions. My purpose here is not to query their conclusions but rather I want to uncover several key elements of the behaviouralist method which are important to a critique of *their* behaviouralist methods. Dahl divided New Haven's citizens into two strata (Dahl, 1961a, 90–100): *homo civicus* and *homo politicus* (Dahl, 1961a, 223–6). The former were politically apathetic and took little part in day-to-day politics but could be roused when they saw their interests threatened. The potential power of *homo civicus*, through the various political channels open to them, kept the leaders in check (Dahl, 1961a, 93). *Homo politicus* were those few who engaged regularly and actively in the politics of the community. *Homo civicus* were assumed to be apathetic because they did not engage in politics. They were assumed, merely because they behaved differently from *homo politicus*, to have a different psychology from that of *homo politicus*. Hand in hand with this different psychology went assumptions about the wants and desires of the apathetic citizens. The New Haven researchers assumed that because these citizens did not create political demands through political action they were therefore reasonably happy with the political decisions regarding the issues in their study. However, our theory of action does not allow us to make those assumptions, for individuals spend their resources where they believe the payoffs will bring the greatest marginal utility. For many of *homo civicus* these calculations alone will lead them to avoid political action merely because they do not expect it to bring the greatest rewards (see below).

The behaviouralists felt that the researcher should not bring her own values to bear upon her work. Thus researchers should only study issues in which people had demonstrated through action that they had an interest. Dahl (1961a, 52 fn. 1) writes:

terms such as benefit and reward are intended to refer to subjective, psychological appraisals, by the recipients, rather than appraisals by observers. An action can be said to confer benefits on an individual, in this sense, if he *believes* he has benefited, even though from the point of view of the observers, his belief is false or ethically wrong.

Homo civicus took little part in politics because he saw little to be gained from it. In this *homo civicus* was probably right, but critics of behaviouralism did not want to leave the matter there. They suggested that people do not take part because many issues in which they have a *prima facie* interest get quashed prior to agenda-setting in the narrowly political realm. Bachrach and Baratz coined the phrase 'non-decision' to denote issues that the behaviouralist methods could not encompass. They write (1970, 44):

> A nondecision … is a decision that results in the suppression or thwarting of a latent or manifest challenge to the values and interest of the decision-maker. To be more nearly explicit, [please KMD] nondecision-making is a means by which demands for change on the existing allocation of benefits and privileges in the community can be suffocated before they are even voiced; or kept covert; or killed before they gain access to the relevant decision-making arena; or, failing all these things, maimed or destroyed in the decision-implementing stage of the policy process.

Bachrach and Baratz suggest that there are four forms of non-decision-making: (1) force (including harassment, imprisonment and murder); (2) the threat of sanctions; (3) the use of prevailing norms, rules and procedures to 'squelch' issues; and (4) the reinforcing or creation of new norms or values to crush incipient conflict. In the same context they mention individuals' decisions not to press for demands because they feel they are bound to lose. The costs are greater than the benefits.

Thus it seems that a non-decision is a decision all right – but one which the behaviouralists did not study. Polsby (1979) rightly complains that in their own study of the race issue in Baltimore, Bachrach and Baratz seem to do little that Dahl, Wolfinger and Polsby himself did not do in New Haven, indeed Bachrach and Baratz do rather less. But what we might argue is that the sorts of issue we choose to study will determine our views upon individual powers. Thus the decision of the behaviouralists to study issues in the narrowly political realm was tainted with their own value judgements about individual interests. The race riots in New Haven just a few years later (see Ricci, 1971, 158) and the question of Yale's low tax burden in New Haven provide evidence of the existence of issues perhaps wrongly excluded from the study of Dahl and his associates. Some analysts suggest that the evidence provided by the behaviouralists themselves do not sustain their conclusions (Morriss, 1972; Domhoff, 1978). But this is a critique of the pluralist conclusions of the New Haven study rather than a critique of the behavioural method.

A critique of behaviouralism as such is suggested by an issue 'related' to non-decision-making. The 'rule of anticipated reactions' (Friedrich, 1941) suggests that people may not press their demands because they feel it will not get them anywhere. Behaviouralists suggested that we just cannot study these 'non-events' (Wolfinger, 1971a) and Bachrach and Baratz concurred (1970, 40). The whole issue of 'non-events' and the debate it created (Merrelman, 1968; Frey, 1971; Wolfinger, 1971a, 1971b; Parry and Morriss, 1974; Debnam, 1975) is a blind alley which is actually totally irrelevant to what Bachrach and Baratz were writing about. It seems to have developed purely because of the absurdity of calling a decision of a certain sort a 'non-decision' and should stand as a reminder to all political scientists of the power of bestowing titles upon our concepts and a warning to write more clearly and less technically than we are apt to do.[4]

In fact we do require a means of recognizing anticipated reactions which I will attempt to provide later (Chapters 3 and 4). But some behaviouralist critics wish to go further. Connolly (1983) and Lukes (1974) suggest that individuals may have 'real' interests which are contrary to their own perceived interests. Perceived interests or 'desires' are the products of the political system which may go against their real interests. Such views are anathema to the behaviouralists. Polsby (1980, 224) writes:

> A definition relying upon 'enlightenment' in some form or other amounts to an argument that actors would choose differently if they knew what analysts knew; in short, it consists of a substitution of analysts' choices for actors' choices. The difficulty with this sort of definition is not simply that in some cases analysts are less able than actors to discern what the correct choices should be from the actors' own perspective. Rather, the problem is that there is no method for determining when analysts are choosing better than actors in [sic] the actors' behalf and when they are not. It is not always clear, for example, that long-run benefits are to be preferred to short-run benefits, or vice versa, although analysts and actors may choose quite differently between these alternatives.

That there are no methods is false; I demonstrate a method in the next chapter. What is true is that analysts are as prone to error as everyone else. However, this epistemological problem stops only the sceptic from conducting research. The psychological assumptions governing *homo civicus* are no better founded than contrary assumptions. Consider the simple Prisoners' Dilemma:

		I_j	
		C	D
I_i	C	R, R	S, T
	D	T, S	P, P

The two players have two possible courses of action. They either do C (cooperate with each other) or they do D (defect from cooperating with each other). There are four possible outcomes represented by the payoffs (T, R, P and S) to each player in the four boxes of the matrix. Each player orders the outcomes ($T > R > P > S$). Thus each player would sooner defect from cooperating whilst the other cooperates, defect if the other defects, but would prefer mutual cooperation to mutual defection. The D strategy *dominates* the C strategy since both players prefer to play D *no matter what the other player does*. Yet when both play their dominant strategy, D, both end up with their third-most-preferred outcome, P, whereas if they had both played their dominated strategy C, they would both have got their second-most-preferred outcome, R. Universal cooperation is universally preferred to universal defection, but universal cooperation is individually unstable and individually inaccessible.

The player who plays D and ends up with P still prefers R. Thus playing D cannot *on its own* tell us what I_i prefers; we first need to understand the structure of the choice situation. It is not out of order to argue that, had the choice situation been different, say an iterated Prisoners' Dilemma, the individuals would have chosen C rather than D. The method which is suggested in the next chapter to overcome Polsby's problem trades upon this point.

Behaviouralism, then, is the study of politics by looking at the behaviour of people. However, studying behaviour requires an explicit theory of action and an understanding of the structure of the choice situations within which individuals make their choices. Rational choice modelling is the best way of attempting this process. The methodological problem for this behaviouralism is to demonstrate that individuals may not act in ways which are in their own best interests. In Chapter 3, I will try to make sense of the concept of 'objective interests' which are consistent with a form of behaviouralism.

3

Preferences and
Objective Interests

3.1 'Subjective' and 'objective' interests

Political power should not be equated with getting what one wants, nor lack of power with not getting what one wants. Preferences and power are intimately linked though, for we assume that people do generally act in order to promote their wants. However individual wants are not merely givens which themselves require no explanation. Actions are explained by examining beliefs and desires. Our beliefs derive from what we see around us and whilst simple desires may be sociologically inexplicable – some people prefer strawberries to raspberries and some raspberries to strawberries – other desires develop from our beliefs. These may be said to be complex desires. This leads us to a maxim of sociological explanation: simple beliefs and desires require simple explanations, complex beliefs and desires require complex explanations. A reordering of preferences may occur as a result of new information which changes our belief set about the options in our preference schedule. We may study the information received by individuals in order to understand why their preference schedule alters. Thus, we may say that individual preference schedules are relatively enduring.

The relationship between preferences and interests is not an easy one to specify. If we equate interests with the highest ranked options in a preference ordering then we can never claim that individuals may be unaware of their own best interests. Whilst some analysts have been happy to accept this conclusion few of us actually recognize this as a truth in our own lives, or we would have no reason ever to re-examine our preferences. There is, then, an obvious empirical sense in which objective interests exist; specifying their nature is much harder and the concept of 'objective interests' cannot be used in a cavalier manner. Lukes (1974, 34) suggests that there are three ways of dealing with interests. First, the 'liberal' view, which

> takes men as they are and applies want-regarding principles to them, relating their interests to what they actually want or prefer, to their policy preferences as manifested by their political participation.

Second, the 'reformist' position, which sees that not all wants and preferences are revealed by the political system; but, whilst it still equates interests with preferences, it

> allows that this may be revealed in more indirect and sub-political ways – in the form of deflected, submerged or concealed wants and preferences.

And finally, the 'radical' view, which

> maintains that men's wants may themselves be a product of a system which works against their interests, and, in such cases, relates the latter to what they would want and prefer were they able to make the choice.

The distinction between the three, whilst partly normative, is largely methodological. The need for 'objective' interests according to the latter view arises because the wants of individuals are created by forces outside them. If those forces were different then people would have different sets of wants. 'Objective interests' in this view are either the wants we would have without the intervention of outside forces, or those wants a rational individual would choose (if wants can be 'chosen') taking into account all possible outside forces.

There are many criticisms of such a view. First, can we make sense of the perfectly rational person who must choose between different sets of wants? Surely all wants are at least partly a product of the society in which we live and so there can be no totally disinterested 'Archimedian point' from which to make such a judgement (Lukes, 1977, 187–90). Second, given the impossibility of such a disinterested position, any attempt to criticize the revealed preferences of individuals will simply introduce the analyst's own preferences (Polsby, 1980). Third, the rational choice approach itself assumes that individual preferences are *exogenous* to the model (or endogenous to the individual). This means that the modeller just assumes that individuals have certain preferences and is not concerned where they come from. The model does not explain preferences; it explains outcomes, given those preferences. Individuals are said to have moral sovereignty over their own desires or privileged access to their own best interests (Buchanan, 1977).

The first two points are undeniable but do they lead to the total rejection of any account of the individual's mis-specification of her own interests? The third is the approach which we wish to adopt to provide a rational choice account of political power. However, as we have already noted, we cannot just accept the exogeny of preferences in a work on power, so we do need to examine the cause of them to some extent. Take a simple example. It would be wrong of me to suggest that anyone who claimed they preferred the taste of white bread to that of brown bread was wrong because in fact they must *prefer* that of brown bread as it is nicer. However, I might be right to claim that if you were to eat brown bread over a long period of time you would come to prefer it to white; but that is a prediction which may be falsified. I might also be right in claiming that you prefer white bread because that is what you ate as a child; and might also be right to argue that your parents fed you white bread because it was thought (until recently) to be superior, owing to nineteenth-century social snobbery (Tannahill, 1975, 316). Thus some sort of power relations may sometimes be worked into an account of a simple taste preference. But there is a point at which people simply have certain beliefs and certain desires which cannot be usefully further explained. They prefer x to y because they do and that is that. This is the most trivial explanation of preferences that is possible. We may refer to the maxim of sociological explanation suggested above.

We require a simple strategy for understanding individual interests: one which accepts that people just have certain desires and that is that, but which allows for the possibility that individuals may misunderstand some of their own interests. The fact that we recognize mistakes in the past leads us to recognize that we may make mistakes about our own interests today. An epistemological experience leading to an ontological discovery. In other words, individuals have objective interests which may be different from the interests they proclaim for themselves. This strategy must be compatible with behaviouralism and the theory of action described in Chapter 2. It will be explained by utilizing three theses.

1. '*Ontological thesis*' – what is in one's interests is much more than what one merely desires because interests are partially dependent upon needs. Individuals may be wrong about their own interests because they are

unaware of the particular needs concomitant upon their desires. We can thus make sense of 'objective' interests without entirely divorcing those interests from individuals' own deep-rooted desires.

2. '*Epistemological thesis*' – whilst some interests are entirely dependent upon certain desires which may be said just to happen (they are endogenous to an individual), most interests are also dependent upon factors external to the individual (they are exogenous to the individual). If we can know an individual's endogenous desires, we may then be able to discover those interests in precisely the same way the agent does for herself. Thus interests may be said to be 'objective' in a further sense; they are open to inspection by all and individuals do not have privileged access to their own interests.

3. '*Methodological thesis*' – we can discover individuals' interests by studying their behaviour. But the behavioural method must include both a theory of action and knowledge of individuals' choice situation. Careful behaviouralism includes both a theory of action and analysis of individuals' choice situations. However, armed with innocuous assumptions about individuals' desires and with analysis of the conditions under which choice is made, we can discover what individuals' interests are.

These three theses may sustain conclusions rather different from accounts usually associated with behaviouralism. They allow us to maintain an empirical social science whilst recognizing that individuals are not always the best judges of their own interests. We may maintain critiques of early pluralist accounts of the distribution of power in modern society without importing further normative assumptions.

3.2 Need and desire

'Need' has proved to be a controversial topic in political theory because of competing views about what humans need in order to lead a flourishing life. A debate in these terms is bound to be morally loaded and should not be belittled (Lukes, 1977; Taylor, 1985a). There are certain needs – for sustenance, educational attainment or psychological well-being, for example – to be met before individuals may lead the sorts of life they desire and many of these may be demonstrated with careful empirical analysis. The important difference between desire and need is that the latter is a *modal* term and the former is not (White, 1975). This means that individual needs have to be explained in terms of something else, whereas individual desires do not. Brian Barry (1965, 48) points out:

no statement to the effect that x is necessary in order to produce y provides a reason for doing x. Before it can provide such a reason y must be shown to be (or taken to be) a desirable end to pursue.

Sometimes when people claim that they need x it turns out that they only need it for some trivial y. But generally when we claim that someone needs x we are making a strong moral claim, the strength of which depends upon the nature of the y that explains the need. There have been various recent attempts to analyse 'need' non-modally (Miller, 1976; Thomson, 1987). Miller, for example, attempts to make sense of 'need' as something intrinsic rather than instrumental, but he still defines needs as those things which are *necessary* or *essential* for each individual to pursue his life-plan (Miller, 1976, 134). Thomson (1987) ends by admitting that individual needs are defined in terms of what harms individuals' essential nature.

Thus needs can only be defined modally and this has important consequences. The desire to produce certain ends may create in individuals needs of which they are unaware. This is because their need is a relation between their desired end and their present position, but they may not realize what is necessary to bring about their desire.

A further and deeper difference between desires and needs arises through the *intensionality* of the mental: that is, what we desire is always something under a description. If an individual needs an object under some particular description of it, it does not follow that he needs a particular object, just one which falls under *that* description. But, whilst one can desire something under one description but not under another, one cannot need something under one description rather than another.

> To want to kill the man who is blocking your escape does not [entail] wanting to kill your own son, even though it is your son who is blocking your escape. What one needs, on the other hand, one needs whatever its description. If in order to escape, one needs to kill the man blocking one's way and the man is one's own son, then one needs to kill one's own son. (White, 1975, 112)

Whilst I may have my own (perhaps peculiar) reasons for desiring something, I cannot have my own reasons for needing something. Those reasons are general (objective) reasons and not individual (subjective) ones. However, the assignation of utility or value to something depends upon the way we understand it – we value it under some particular descriptions and not others (Schick, 1982). Any ascription of desires entails, given the extant circumstances, a set of needs. These needs hold, no matter what the individuals recognize about them. I desire to escape and hence need to kill the man blocking my escape. I may desire not to kill my son who happens to be the man blocking my escape. Two desires here happen to conflict. Once that conflict is recognized, one desire may take precedence over the other. If, however, the two desires are incompatible then rational individuals must give one up. If the desires are practically impossible to combine then again rational individuals should give one of them up. Of course, we may wish for something we recognize to be impossible, but wishes are not desires.

The process of forming desires is part and parcel of our experience in which our beliefs play an important role, and ideally our beliefs form a consistent set. Our desires are not clearly formulated prior to our experience of the environment, and during our experience of it they change. A recognition of the needs of our desires may well change them, but the desires and their needs should be carefully separated in analysis.

3.3 The 'ontological thesis'

The ontological thesis requires the disjunction between human 'needs' and human 'desires' provided by the modal nature of 'need' and the non-modal nature of 'desire'. Given the relationship between desires and needs we can see that views which equate interests with desires are false. If I believe that nuclear power is dangerous because the scientists tell me so and expensive because the economists tell me so, then I may well calculate that it is in my interests for Britain to have coal- or oil-fired power stations rather than nuclear ones. But if the scientists and economists are wrong and nuclear power is safe and inexpensive in the long run, then I may well make a different calculation. This failure on the part of individuals to understand their own interests is a failure in belief acquisition and/or mode of reasoning. It does not suggest that individuals may make errors in their judgements about intrinsic value. It could only lead to this further error in so far as judgements about intrinsic value depend upon individuals' beliefs about the world and their modes of reasoning.

I take this to be an innocuous conclusion. After all, we are all, bar the most bigoted, willing to admit that some of our beliefs are wrong, even if we do not know which ones. In the 'preface paradox', the philosopher writes in the preface to her book, that, whilst she believes that each of her conclusions is true, she knows that some are false.

So there are at least two ways I may be wrong in believing that some course of action is in my interests. First, I may be right in believing that I need u in order to get the wanted x, but be unaware that u also leads to the consequence y which is feared more than x is wanted. (If u is truly necessary for x to be satisfied then this is the same as saying that I want x under one description but not under another.) Second, I may want x yet be unaware that I need u in order to get it and so oppose u despite my want for x outweighing my opposition to u.

Thus the 'ontological thesis' – individuals may be wrong about their own interests because interests are dependent upon the needs which spring from one's own desires. Individuals may be unaware of these needs and hence unaware of their own interests. Analysis of interests in terms of needs gives an objective account of interests.

3.4 The 'epistemological thesis'

How do individuals develop desires? For 'liberal' analyses of interests this does not matter. That individuals have them is good enough. We proceed from there. For the purposes of political analysis the 'liberal' approach is a good start, from which we may provide a reflexive account that says more about want-generation.

We may begin with an ostensive categorical distinction. There are two sorts of interest: 'endogenous' and 'exogenous' (Dunleavy, 1988). A simple way of explaining the distinction suggests that endogenous interests are ones that we just have. We like apple pie and therefore have an interest in the production of good-quality apple pies. We can say we have our endogenous interests by virtue of our personal characters (however those characters are developed). An exogenous interest is one which we have in virtue of the circumstances in which we find ourselves. A coal miner has an interest in the wages and work conditions of coal miners, a shop assistant in the wages and work conditions of shop assistants. So do the owners or managers of coal mines and shops, though of course the two sets of interests may be in direct conflict. These interests can be said to be exogenous because person i only has a particular interest in the wages and work conditions of coal mines because he is a miner. If i was a shop assistant then, whilst *he* might still desire that coal miners be well paid and work under safe conditions, he has a more particular interest in the wages and working conditions of shop assistants. And the interests of shop assistants and coal miners may come into conflict. Person i would also have different interests if he were the owner of the coal mine. He would have these different interests whilst remaining unaltered under every other description. This analysis does not *entail* that interests are structurally determined, though it does *imply* that interests are structurally suggested. That is, when deciding what our interests are, the most important considerations are those aspects of the reality around us which affect us most deeply.

We do not need to make any particular assumptions about individual desires in this thesis. A careful analysis of a particular individual's behaviour, given his beliefs, can allow us to understand those desires, and to criticize them if they are based upon false belief, or to criticize that person's actions if they are not the best means to his or her ends. Given the actions and wants, we can discover the individual's belief set, and likewise criticize that set if it does not suit the reality around that person. However, we do need to make particular assumptions about underlying desires under the methodological thesis. A model explaining the actions of individuals in the above examples might assume that *ceteris paribus* individuals desire higher rather than lower wages for themselves and better rather than worse work conditions. They assume that *ceteris paribus* owners of factors of production prefer higher rather than lower profits, and so far as this is achieved by paying lower wages and having poorer working conditions for employees then that is what they desire. But I take these assumptions to be innocuous. The analysis only requires the central assumption of neo-classical economic theory, viz. rational individuals allocate resources in order to maximize their marginal utility. Different models can operate with different types of standard economic assumption: self-interest, altruism or a mixture of both (Margolis, 1982). It can assume that people are more willing to help their family, friends, workmates or countrymen than aliens. It can use any set of assumptions about personal desires to argue that individuals have interests externally defined by the position which they occupy in the social system. Having said that, however, we may not wish to hold that desires may vary without restriction. Barry and Rae (1975, 382) suggest that interest

> always appears to have carried an emphasis on material advantage and thus to find its home especially in economic and quasi-economic discourse.

Thus we may not want to say that it is in someone's interests to have food sent to Ethiopia, even though she is morally committed to famine relief.

The ostensive categorical distinction is what it states. It is an ostensive definition of two categories which makes a distinction which may be useful for social analysis. But the distinction is hard to maintain as a natural category. For example, the anti-nuclear movement may be treated as an endogenous group. For some purposes that is adequate; people do decide whether or not they think nuclear power is a good or bad thing. The decision to join or leave the group set is a decision. In Hirschman's (1970) terms, once one has decided to leave exit is easy, indeed the decision to leave is exit from the identity set. But of course individuals do not make important decisions like these in a vacuum. They make decisions within the social situation in which they find themselves. We would be likely to define as exogenous an individual's interests in relation to nuclear power if he worked at a nuclear power station. But I think the distinction is useful for social scientific research, as I shall show when discussing the 'methodological thesis'.

This analysis has an important epistemological point. The considerations which enter into an individual's own calculations of her best interests are precisely the same considerations which enter into anyone else's calculations of that individual's best interests. If we make the correct assumptions about the individual's desires (which we may check by asking her) and we know that person's situation as well as she does, then we can make the same calculation. If we know the situation better than she, then we can calculate those interests better than she can. Thus we may answer Polsby's criticism cited on pages 29–30. Of course he is right that analysts are as prone to error as the people they study, but I reiterate that that stops only the sceptic from carrying out research.

This conclusion may be controversial. But it is also incontrovertible. In order to dispute it the critic must argue against the modal status of 'need' and/or against the fact that individuals make decisions in the light of the reality which they perceive around them. That reality may also be perceived, sometimes more clearly, by the analyst. The desire to dispute the conclusion proceeds from the natural fear that it somehow justifies authoritarianism. It does not. It might perhaps be used as an element in an attempted justification. But this is not good enough a reason to reject it. There are a host of good reasons against a purported justification of this kind which allow us to reject such a 'Maginot Line strategy' (Taylor, 1985b, 217). I will simply state that at times it may be right to force people to do something against their will because it is in their best interests, but that generally speaking it is not. In political situations the decision as to which those right times are will always be controversial. However concern over moral questions of policy formation should not blind us to good social scientific methodology. The argument above justifies a behavioural approach to political analysis, yet it questions the way some behaviouralists have gone about their work. I will spell out a 'methodological thesis' which follows from the ontological and epistemological ones.

3.5 The 'methodological thesis'

Lukes believes that his first definition of interest is required for behaviouralist approaches to political analysis. However, this is only so if we equate interest and action. Such an equation is most clearly seen in the work of Bentley (1967), but also emerges less explicitly in many modern analyses. A fairly circumspect analysis of interest in the liberal tradition is given by Nelson Polsby (1980, 225) when he says:

> It is … compatible with other pluralist beliefs to assert, at least *a priori*, that what an individual or group wants, or what they say they want, or in some way indicate they want is, by definition, what their political interests are.

Polsby is aware that this definition makes it difficult to analyse the claim that someone acted against their own best interests. So, he suggests two supplementary axioms: (1) the relationships between the causes and effects are frequently well and generally understood in a large number of situations in which human beings are faced with choices that have direct consequences for their overall value positions; (2) people generally try to maintain or improve their overall value positions. When they choose not to do so in situations where the causes and effects are clearly specified and well known, they can be said to be acting against their own interests.

But the supplementary axioms will not save Polsby. Consider again the simple two-person Prisoners' Dilemma.

Two-Person Prisoners' Dilemma

$$I_j$$

		C	D
I_i	C	R, R	S, T
	D	T, S	P, P

where $I_{ij} \{T > R > P > S\}$.

Each actor gets their third-most-preferred outcome by attempting to maximize their payoff by following the *dominant* strategy (*D*); whereas each could have attained their second-most-preferred outcome by following the *dominated* strategy (*C*). (*D* is the dominant strategy because no matter what the individual *j* does it is always in the interests of individual *i* to do *D*.) Even given Harsanyi's (1977, 113–18) strong rationality postulates, (a) *maximization* (individuals choose strategies to maximize payoffs), and (b) *rational expectations* (each expects the other to use strategies to maximize payoffs), individuals will choose the dominant strategy. There is no question that the causes and effects are not well understood, nor that they have direct consequences for individual value positions. There is no question that each individual acts rationally to promote their best interests, but there is also no question that those interests would be better served if each were to follow a different course. Each individual faces a situation which structurally suggests a course of action, but nothing in that action *alone* can tell the analyst which cell of the matrix each individual prefers. We need also to understand the structure of the decision situation.

In effect Polsby's axioms ignore the collective action problem (popularized by Olson, 1971). Behaviouralist methods ignore the fact that an agent's interests constitute part but not all of her reasons for a particular action. The agent must also consider the possibilities open to her. These possibilities will depend not only upon the *way* she assesses them but also upon the *actual means* by which the action is available to her. Thus such behaviouralist methods operate without a good theory of action.

It may be objected that individuals' interests *are* revealed by action since the action is the practical conclusion of a rational calculation: the pros and cons of possible alternative courses of action are weighed up and the course chosen reveals what that individual believes to be best for herself. It is true that 'ideally' she may prefer *a* to *b*, but does not act so as to help bring about *a* because the costs of that action outweigh *a*'s extra value over *b*. Thus the individual calculates that the inaction which helps bring about *b* is in her interest. So we could argue that the passivity of blacks in Baltimore in 1964 (Bachrach and Baratz, 1970) was in their interests as individuals. But such an argument hides a number of highly dubious assumptions.

First, it assumes that the situation of the person making the practical deliberation is neutral regarding those interests. It does not allow for the possibility that it is not in the individual's interests to be in the conditions under which the practical deliberation is made. Where those conditions have been deliberately created by others then we have a clear example of power. I may be about to join the union but decide not to when I am told that if I do I will lose my company house (Gaventa, 1980, 96–9). Second, it ignores freeriding. The individual may calculate that the chances of being pivotal in bringing about *x* are remote, and hence the expected extra benefits consequent upon doing action *a* to help bring about *x* are less than the expected costs of doing *a*. If it turned out that she was pivotal she would do *a*. (Schelling, 1982 and Taylor, 1987 describe conditions under which such calculations are made.) Third, it assumes that, if two people (A and B) value some outcome *x*, and A is prepared to pay $10 for it but B only $5, then A values *x* twice as much as B. But this is only true if A and B have exactly the same disposable income. A sum of $10 may be chicken-feed to A, but $5 the daily income of B. We must not assume that individual monetary value assignations are any use for interpersonal comparison of individual utility calculations. It is relative monetary costs that are important (Goodin and Dryzek, 1981).

If we assume that wants as revealed by actions exhaust individual interests, then we cannot allow for the possibility that individuals are unable to articulate interests which are, or are believed to be, outside their feasible sets. Individuals may choose one course of action rather than another simply because they prefer its results. But often some courses of action are preferred over others because the interests they promote are either (a) the most 'immediate': that is, the most easily perceived, or (b) the most easily achieved. For example, the interests of groups of workers are often more immediate and perceivable than the interests of consumers. But all producers are also consumers. Individuals may also choose strategies to achieve a less-preferred outcome instead of strategies to achieve a more-preferred outcome if the former have a much higher probability of succeeding. Such choices depend upon how much one option is preferred to the other, the probability of attaining each, and how risk-averse is the individual.

In fact Polsby (1980, 217) objects to Crenson's (1971) study of the air pollution issue in Gary and East Chicago on similar grounds. He says:

> Many people trade off air pollution against employment, and it is not necessarily the case that they do so unwittingly. Therefore merely showing that in some communities a given issue is raised and tends to lead to a particular conclusion does not discriminate between the possibility in other communities that an issue is being suppressed and the possibility that a genuine consensus exists although that consensus is contrary to what an observer can find elsewhere.

In Gary individuals may well have felt that pursuing clean air policies would lower their employment prospects. But this does not show that they had no interest in having clean air: it merely shows that they had a greater interest in employment. It also says nothing about the power relations between the steel employers and the townsfolk. If we recognize the distinction between 'needs' and 'desires' specified above, then we do not need to impute a power relation between the townsfolk and the steel company. (Though that is not to say there is not one, see Chapter 5.) Crenson notes that the individuals in Gary do not carry out a certain action *a*, which is inconsistent with what he believes their interests to be. This may be explained by the fact that they have other, *prima facie* conflicting, interests which override the interests that Crenson imputes. They do b in order to get y (employment); in order to get z (clean air) they need to do a. But so far as the individuals are concerned, doing b implies doing not-a. Thus they must choose between ends y and z. But doing not-a does not mean that they have no interest in z: only that that interest has been overridden by y. However, there may be a state of affairs in which y and z are compatible. In order to achieve this, they need u. In this example what are needed are pollution controls throughout the United States, so that no community is relatively disadvantaged by them. If the United States steel industry then suffers from competition abroad, world-wide ordinances are required. These are just larger and larger collective action problems. Interests claims must always be made within the widest context.

If the conflict is truly a collective action problem then u may be attainable. Individuals may not have realized that they need u, or if they have realized it they may believe that the costs are prohibitive. Lack of collective organization affects their power, but we do not need to impute power to the steel owners to understand this. Schattschneider (1960, 71, emphasis added, original all italicized) wrote 'organization is the mobilization of bias'. It is lack of organization that affects the power of individuals.

In collective action problems certain needs are requisite for certain of our desires. In order for collective action to be taken, some conditions need to be fulfilled whether or not any individuals want those conditions to be fulfilled. Those conditions might include the lobbying power of an effective group organization, or the workings of a bureaucracy, or higher taxation, or greater competition in a particular market. If certain requirements are necessary to overcome collective action problems, then we must meet them whether or not we 'ideally' want them. The problem is that the ideal world is usually not a possible one.

The individual costs of acting to promote policies that are in our interests may *be* greater than the benefits we expect from their promotion. In fact, they may be so great that we do not even think such interests are feasible. In order for them to be promoted or made feasible we need to move to a situation in which the costs are less, and where the outcome appears feasible. Our wants may thus be less restricted. We can begin rationally to desire outcomes that before seemed impossible. In this way we are able to sustain the heart of Lukes's 'radical' definition of interests within a behaviouralist framework to political analysis. For, insofar as the choice situations in which individuals find themselves restrict the feasible set, they may be said to work against their wider interests. All that we require for such a fusion is the realization that we cannot study human behaviour without a theory of action and without the understanding that situations channel individual action just as collective actions create situations.

I left aside the question of the normative implications of the ontological and epistemological theses above. The normative implication of the methodological thesis is similarly worrying to the 'liberal'. If we admit that individuals may not be the best judge of their own interests, how do we avoid observer bias in analysis of the political system? Observer bias nevertheless enters in the type of questions that any observer poses, and simple observation of action cannot tell us what individuals believe their interests to be, precisely because of the collective action problem. Some interests have to be assumed for the sake of analysis. This is just as true for liberal pluralist studies of the New Haven sort as for any other. All too often the observer bias claim turns into an ontological claim which equates the interests of individuals with the manner in which they act. But the answer to our question is not simple. In any given interest ascription the analyst must justify that particular ascription (Smith, 1981). There is no general formula by which to judge such ascriptions. Each must be examined in the light of the arguments which support them. This is so, of course, because the analyst may mistake the interests of his subjects in precisely the same manner in which the subjects may get them wrong. We all use the same sort of evidence in interest ascription and we are equally subject to the problem of knowledge.

3.6 Interests in preference schedules

We have a clear method within rational choice for considering conflicts of individual interest. The preference schedule relates a set of possible options to an individual in the order in which that person prefers them. However, the simple Prisoners' Dilemma demonstrates that individuals will not necessarily act in such a way as to produce an outcome which is higher in their preference ordering than the outcome of the action they actually take. Given that individual preference schedules do not make much sense outside the constraints under which the schedule is formed there is a problem with making sense of the notion of individual subjective interests in the first place. The concept of individual subjective interests given a preference schedule subject to constraints can only mean the highest option in the preference schedule. But this is not what we ordinarily mean when we talk about an individual's best interests, even when the person doing the talking is the individual herself. Consider once more the structure of preference for a prisoner in a simple Prisoners'

Dilemma. She orders the options: $\{T > R > P > S\}$. Her first preference is T. Is this in her best interests? In a sense it is, for getting off scot-free is obviously more in her interests than spending a year in gaol. Yet we know, given the structure of the simple Prisoners' Dilemma, that the individual's attempt to obtain her most-preferred option is likely to lead to her getting her third option. That choice then is not in her best interests or, rather, being in the situation of having to make that choice is not in her best interests. In other words, it is not in your own best interests to be faced with the decision structure called 'Prisoners' Dilemma'. Can we say that this notion of best interests is also merely a preference? In other words, that individuals merely prefer not to be in Prisoners' Dilemmas and hence their true preference structure is $\{A > T > R > P > S\}$ where A = not face a decision structure called 'Prisoners' Dilemma'? Here we have argued in a circle to best interests with preferences again. However, this cannot be right, for earlier preference schedules were recognized to be the order in which individuals place options given constraints and, by hypothesis, the constraint in this case is that they face the decision situation called 'Prisoners' Dilemma'. In other words, the option A is just not allowed in the preference schedule. This may be one way in which we can distinguish, even if only by definition, the notion of individual interests from mere preferences. Individual interests are constituted by the conditions under which individuals would like to be able to choose from the options under offer. Thus an individual's best interest is being able to choose his first preference, T (that box in the matrix), rather than being able to choose a strategy, D (that line in the matrix). His best interests are unbounded by the actual constraints under which choice is made; his preferences are bounded by the choice constraints.

We can say more. We noted earlier that one way of demarcating objective interests from subjective desires is by recognizing that certain needs are often requisite for certain desires to be satisfied, needs of which the individual may be unaware. Thus the argument that choosing the box T rather than the line D is the individual's best interests requires also the recognition that certain conditions must be satisfied in order for that option to be available. These conditions require going outside the Prisoners' Dilemma structure. Of course, if those conditions are not naturally attainable then the needs cannot be met and the interest vanishes, for one cannot be held to have interests which cannot exist.

More complex calculations can also be made. For example, whilst individuals prefer T to R they also prefer R to P. The conditions for individually attaining T may be too great: the players cannot step outside of the Prisoners' Dilemma. The option T thus becomes practically (though not logically) infeasible. If so, then R may be recognized as being in each individual's best interests. The requirements for attaining R may involve changing the choice structure for each individual from that of the simple Prisoners' Dilemma into something in which the option R is available for all and which may be in each individual's interest. Here, then, we have attained the divergence we were looking for. T is the individual's most- preferred option, but R is in his best interests: for only R is attainable, once the constraints under which choice is made are altered by the collective action of the individuals whose preference schedules we are considering.

We must also recognize that once P has been achieved their preferences are now formed. Had they achieved R rather than P they might have had a different preference schedule, for then they would have seen a different set of options as feasible. Unintended consequences of individual action therefore affect the basic wants and desires which form the initial starting-point in our analysis (see Chapter 7).

None of this requires the assumption that there is an actor who has power and is creating individual wants and desires in order to further her own ends. We can see how individuals may both not reveal their preferences to a casual observer and mis-specify those preferences to themselves without needing to assume that another person is pulling the strings. Such an assumption is an example of the 'blame fallacy'. The 'blame fallacy' asserts that because something did not turn out as intended there must be someone to blame. We can see this fallacy in action every time there is a disaster, whether natural or human. Sometimes the burden of responsibility for a catastrophe may lie with individuals, sometimes not; but there will always be an effort to lay the blame at one doorstep or another. Chapter 5 will develop this point and demonstrate how many scholars studying power have committed the blame fallacy in relation to ascriptions of social power.

4

Political Power and Bargaining Theory

4.1 Distinctions in the definition of power

Like other concepts in social science such as 'freedom' or 'authority', the term 'power' does not make much sense unless used in conjunction with a natural object and the scope of its application is restricted. When we say that person A has power we are specifying a property of A and in order to understand that property we need to add scope modifiers – A has power to do x; or A has power over B to achieve outcome x. The most important distinction between different uses of the term 'power' has been over the understanding of the scope of power claims. Some have seen political power as the production of social outcomes in general, others have seen it primarily as conflict between actors. The distinction is sometimes expressed as the difference between 'power over' and 'power to'. A person may be said to have the power to do x – power to; and a person may be said to have power over B – where B is another person. 'Power over' implies 'power to' of course. If a person has power over another then they have power over B to do (not do) x.

Riker (1969a) makes a similar distinction between 'ego-oriented power' and 'other-oriented power'. Ego-oriented power is 'the ability to increase ego's utility', whilst other-oriented power is 'the ability to decrease alter's utility' (Riker, 1969a, 114). Riker's definitions do not really capture the distinction he wants, since he also describes ego-oriented power as the manipulation of outcomes and other-oriented power as the manipulation of people. But manipulating outcomes to increase my utility may also simultaneously decrease yours, whilst I may manipulate you (for my own ends) in a way in which increases your overall utility. The importance of Riker's distinction is his recognition that for ego-oriented definitions power always exists in social relations. It cannot be eradicated since it refers to outcomes and whenever actors produce outcomes then someone is using power. Note that this does not entail that all outcomes are the result of someone's using power since outcomes often occur which no one intends (see below); and some outcomes may come about outside the influence of any actors.

The distinction between 'power over' and 'power to' may be described as 'outcome power' and 'social power': the first because it is the power to bring about outcomes; the second for it necessarily involves a social relation between at least two people. Both are a species of political power and may be defined:

'outcome power' $\stackrel{def}{=}$ the ability of an actor to bring about or help to bring about outcomes,

'social power' $\stackrel{def}{=}$ the ability of an actor deliberately to change the incentive structure of another actor or actors to bring about, or help bring about outcomes.

The 'incentive structure' specifies the full costs and benefits of attempting to bring about the options in an individual's choice set. Deliberately altering the incentive structure includes adding or subtracting items from that choice set, or changing the relative costs and benefits of pursuing them. The phrase 'changing the incentive structure' is deliberately broad and includes all methods from coercion to persuasion, from overt to covert acts to change costs and benefits, beliefs and (thereby) desires. Such actions may be for good or ill on the part of either actor. But changing another's incentive structure should be both a deliberate act and an act designed to bring about some further outcome. I leave the phrase as broad as possible to get away from interminable terminological disputes attempting to distinguish 'power' from 'influence' from 'authority' from 'persuasion', and so on. Such disputes may be of some interest in certain, particularly normative contexts, but are of less significance to the present largely empirical work.

The use of the term 'deliberate' is stricter with regard to social power than the classic definition of restrictions on negative freedom which includes any action by others closing down my area of choice (Berlin, 1969, xxxvii–ix). 'Deliberate' here means that the aim of the actor has been to restrict the options in the choice set in the way in which they have been restricted. My liberty may well be restricted through the actions of others both by their outcome and their social power. How much moral blame may be attached to their restrictions on my liberty through their actions in these regards is another matter. Depending on the circumstances more blame may be attached to restrictions on my liberty through the use of others' outcome power than through their use of social power. The issues which enter into the moral calculus here include:

1. Social power: their intentions regarding the restriction of my liberty – are they restricting my options for my own benefit (in their calculation)?
2. Social power: if it is a straightforward case of conflict of interests does this justify a power-holder using her power to change my choice situation to attain her ends? Under what situations of conflict of interest would such use of social power be justified?
3. Outcome power: if it is a straightforward case of conflict of interests, how much account should the conflicting parties take of the others' interests? Is a straight battle morally justified, even if those with the greater resources always win?
4. Outcome power: should the power-holder have been aware that her actions were going to affect the interests of others? How far does ignorance of the consequences justify restricting the choice of others?

These are all interesting and important normative questions concerned with human liberty, but ones which I shall not be attempting to answer here.

Obviously the two definitions are intimately linked, for both relate to the ability to bring about outcomes. This is important. It is sometimes suggested that individuals may want to have power over others for no particular purpose. They may merely gain gratification by forcing others to bring about outcomes that would not otherwise have occurred. That may indeed happen, though I suspect far more rarely than cynics might suppose. I will assume that when one actor changes the incentive structure of another she does so in order to bring about or help bring about some particular outcome, though why that actor wants that particular outcome or result will be left open – so it may include cases where the actor desires that outcome merely to prove to herself that she has the power to bring it about.

An important difference between the two definitions is that the first does not necessarily involve more than one actor, whereas the second does, for it specifies a three-place relation. Thus we may talk about Robinson Crusoe, alone on his island, as having outcome power. He can build a shack to live in, a fire for cold nights and make fishing nets to improve his catch. Such outcome power can hardly be described as a form of political power, for politics is necessarily a relation between people and not merely between individuals and their natural environment. But Crusoe's outcome power becomes political when it affects others, even if Crusoe is unaware of those effects. If his shack-building and fire- and net-making have no positive or negative externalities for others then no social relation has been created, but where they do enter others' utility functions then those actions are political ones. For example, if Crusoe's fire creates pollution for the asthmatic Friday, or his efficient fishing reduces Friday's catch, or his shack provides a handy windbreak for shivering Friday, then Crusoe's outcome power may become political. In other words, any action which may cause conflict should the affected individuals or groups realize that that action affects their preferences or interests is a political act. Weber (1978, 1399) says in his most pithy definition of the political: 'Politics means conflict.' This does not mean, I take it, that all conflict is political, but all politics is conflictual and therefore any potentially conflictual act is potentially political (if other conditions obtain). Since conflict may arise whenever preference orderings differ in some regard, virtually any act is potentially a political one. This is a very broad definition of the 'political' which I will not further defend here, but I believe that narrower conceptions of the political fall prey to internal inconsistencies and paradoxes.

Not all of our acts are political ones, for they do not all affect others in the relevant manner, though they may all be replaced with acts which affect others in the relevant manner. In so far as 'acts of omission' carry moral blame as well as 'acts of commission' then failing to act may be political. Thus doing one act which, *ceteris paribus*, is apolitical, may be said to be political in so far as it replaces an act which is political. How far acts of omission carry moral blame will not be pursued here. Furthermore, acts may affect others even if we are not aware of it. As the world becomes more interdependent, with greater numbers of people and ever scarcer resources, then more of our acts become political ones even though our intentions have not altered. The realm of the political is ever-widening and that of the truly private shrinking, though not – we hope – to oblivion. Thus outcome power as a species of political power does involve more than one person, though only the power-holder may herself be denoted in a description of it. However, even outcome power requires scope modifiers which specify what the *person* has the power to do. By studying the scope modifiers we may see how others become affected.

Outcome power must not be confused with freedom, although the two are closely related. Individuals may be free to do an action which they do not have the power to do, or they may perform actions which they are not free to do. Liberty is not the same as ability (Day, 1987). When we talk of legal freedom we are speaking of what we may legally do without sanction from the state. The conscientious objector has the outcome power to refuse to fight but is not free not to fight. Incentives are created to try to make conscientious objectors fight: they may face ridicule or threats from their friends and neighbours, and the state may threaten them with gaol. We may say that conscientious objectors are not free to refuse to fight without facing ridicule and threats and without going to gaol. In this way, by carefully specifying scope modifiers, outcome power and freedom seem equivalent, but it would be ridiculous to claim that the Romanian people were free to overthrow Nicolae Ceausescu because they did so, and that, because they could have done it several years earlier, they had had that freedom for some time. Legal freedom is thus easy to specify. Social freedom is less so, for one's outcome power is affected by the sorts of incentives that actors other than the state set up.

Freedom differs from outcome power in other ways. One may be free to do what one does not have the ability to do. I may be free to be a bricklayer, but I do not have the ability to be a bricklayer because of my arthritic hands. This is not merely legal freedom again for my inability is not caused by anyone's preventing me from being a bricklayer, nor even by anyone's altering my incentives to become one. My inability is not due to the actions of anyone else; I have not been deprived of my freedom to choose a career in bricklaying. Individual outcome power is thus closely related to freedom but is not equivalent to it.

Peter Morris (1987) has tried to distinguish between ability and ableness in order to allow us to say that individual ablenesses are equivalent to individual freedoms. The distinction is based upon the *Oxford English Dictionary*'s definition of 'ability' and 'able':

> *Ability*:
> 2. The quality *in an agent* which makes an action possible; suitable or sufficient power (generally); faculty, capacity (to do or of something).

> *Able*:
> 4. Having the qualifications for, and means of, doing anything; having sufficient power (of whatever kind is needed); in *such a position* that the thing is possible for one; qualified, competent, capable. (Cited in Morriss, 1987, 81, his emphasis)

For Morriss abilities are the capacities we have which we may use under particular conditions (power in a generic sense). Ablenesses are the abilities when those particular conditions obtain (power in a particular sense). However, the particular distinction Morriss makes breaks down under closer analysis unless abilities are equated with genetic abilities or 'internal' resources and ablenesses with 'external' resources (Dowding, 1990). The distinction which exists between the two, and is contained in the dictionary quotation, is a normative one, as Morriss's own example makes clear:

> The rich are able to feed off caviar and champagne; the poor have to restrict themselves to beer and pickles, and are unable to eat more expensive food. This is not due to any lack of masticatory ability on their part, but because of the social and economic environment they inhabit. They are unable to eat caviar, whilst having the ability to do so. (Morriss, 1987, 81)

The reference to the social and economic environment the poor inhabit, which deprives them of ableness, gives the normative rather than the logical sense of ableness. The poor do have the ableness if they break the law – steal caviar or stage a revolution after which it is equally rationed. They may not be able to steal caviar if it is too closely guarded, or if the potential costs of revolution are greater than its potential benefits they thus remain unable to eat caviar. But in these cases their lack of ableness is based upon their abilities, though not their masticatory ones. Thus the distinction between the two is a normative one which distinguishes freedom from outcome power and not the positive distinction between two sorts of power which Morriss tries to capture. The exact nature of that relationship is complex and I will not attempt to specify it here (see Dowding, 1990).

Social power necessarily involves two or more people, for it is a three-place relation. Moreover, social power is exercised when the incentive structure of other actors is deliberately changed to bring about or help to bring about some outcome. This clause is required for both analytic and normative reasons. If the clause were not included then the two parts of the definition of political power would collapse into one whenever outcome power affected others' utility functions. Generally speaking, when we bring about some outcome we thereby affect the incentive structure of other actors. We may make it easier for them to accomplish their aims, or harder, but by causing some change in the environment we have affected others' choice situations. The two definitions are distinct, for in the first some outcome is brought about because the actor desires it and any change in others' choice situation is a by-product and irrelevant to the actor's scheme. Under the second the choice situation of others is changed *in order* to bring about some outcome.

The two parts of the definition of political power are analytically distinct: that is, they are usefully distinguished for the purposes of analysis, though ontologically – even causally – they may be hard to separate. However, I believe the analytic distinction is useful for studying power in society, for it suggests a two-fold strategy. First, study the actors' *choice situation*: examine the status quo to see what individuals and groups can and cannot achieve; or more precisely to see what they believe they may be able to achieve, through what means and with what probability of success. This process itself involves two stages: (a) we may assume that an individual or group faces a parametric environment: that is, we study the difficulties facing the group collectively organizing to achieve common aims without any opposition; (b) then we assume that the group faces a strategic environment: that is, the difficulties of collectively acting are compounded by other actors. (Both of these elements are contained in the Harsanyi bargaining approach, which will be discussed in Sections 4.4 and 4.5.) It is worth while examining both of these elements of collective action without rushing to examine the second only, for otherwise we will miss the lack of outcome power of many groups which apparently face no opposition. This might be the case, though, because a group poses no threat, and there is therefore no point putative opposition organizing. One of the reasons why the group poses no threat is that the members are unable to overcome their collective action problem.

Secondly, we may study the development of the choice situation. How did the actors under study find themselves in this situation? How did the structure of society evolve? How deliberate were the stages of its evolution? How much is the structure a by-product of other aims of the causal agents?

The first stage is ahistorical in that it considers the situation of actors at a given moment without considering how that situation arose. Historians sometimes call this the 'snapshot technique'. The second stage examines the history which created that situation. Sometimes we may be more interested in the first stage than the second; sometimes the other way round. Both stages are required for a fuller understanding of the powers of actors.

Some theorists believe that outcome power is the most important sense of 'political power' (Mills, 1956; Morriss, 1987), most others that social power is basic (Dahl, 1956; Lukes, 1974; Nagel, 1975; Barry, 1980, 1988; Oppenheim, 1981; Therborn, 1982; Connolly, 1983), but basically it does not matter so long as the two uses are distinguished. It may well be that power to bring about outcomes is not the sort of political power that is ordinarily meant when we talk of the power structure. Rather we mean the power to bring about outcomes despite resistance, but to define away outcome power here and ignore its effects upon individual beliefs would be to beg some important questions. The two-part definition of political power – involving both outcome and social power – does away with this problem without begging the question the other way; that is, defining one's power as getting what one wants.

The two parts of the definition need to be kept apart, for there would otherwise be little need for a rational choice account of *power*. If we give a resource-based account of outcomes in which the strength of competing forces over outcomes is given by a set of figures in an equation there would be no need for the use of the term 'power' which could be replaced by the equation itself. In other words, if we equate the power of actors with the resources that they have at their disposal then we no longer require the term 'power'. All we require is the resources which they have in comparison to the resources of all the other relevant actors. (It has been suggested that this is what has happened to the notion of 'economic power' in neo-classical economics: Bartlett (1989).)

If the definition of outcome power were enough to specify political power then the term 'power' could drop out of use. But outcome power does not capture many of the intuitions we have about political power and which theorists have tried to encapsulate in their formulations. These intuitions are ones which concern desire-generation and the way in which our preferences are structured. They are in part determined (or structurally suggested) by the very choice situations we face. We discount certain possible desires or discount some preferences within a possible preference schedule because of our calculations about their feasibility. This factor may lead our calculations to be deliberately manipulated by others for their own ends. Whilst it is true that their power to do so is based upon their resources, this deliberate use of our own preferences should be distinguished from mere conflict over outcomes. A simple resource-based account will lose this manipulation by ignoring the origin of those preferences. The simple resource-based account also stumbles over the stubbornness problem discussed in Section 4.4.

Summary of Section 4.1

The account of political power elucidated here may be summarized as a set of propositions:

(a) Outcome power is the ability of an actor to bring about or help to bring about outcomes.

(b) Social power is the ability of an actor deliberately to change the incentive structures of another actor or actors to bring about or help to bring about outcomes.

(c) Therefore social power is a subset of outcome power and that subset which is most easily identifiable as a form of political power.

(d) Any act which is potentially conflictual is potentially political.

(e) Given an unlimited set of possible individual preferences any act is potentially conflictual.

(f) Therefore any act is potentially political.

(g) Politics has been defined in terms of conflict, but the exercise of both outcome power and social power may not necessarily cause conflict; thus it seems that both outcome and social power need not be exercises of political power. But given the very broad scope of the potential for some act's being considered political on account of its potentially conflictual properties it is unlikely that any act of social power will be considered non-political and many exercises of outcome power will also be political ones.

(h) Liberty is not the same as ability. A person's freedom (where the definition is not modified, for example 'legal freedom') lies in the absence of anyone else deliberately affecting her incentive structure. Liberty is therefore the absence of social power over one. (In any ordinary usage appropriate scope modifiers would need to be added; no actor is totally free *simpliciter*.)

(i) Therefore one can be free to perform some act without having the outcome power to perform that act.

(j) If we concentrated entirely on actors' abilities to bring about outcomes then the notion of 'power' could be dispensed with. All we would need to do would be to look at individual resources and study outcomes.

(k) Therefore we need to study the ways in which actors seek to change the incentives facing others (social power) and the ways in which actors unintentionally affect others. These are usually called 'externalities', or could be referred to as 'non-social power'.

In the previous chapter I considered the nature of preference schedules and their relationship to individual interests. In Chapter 7, I will reconsider preference formation in the light of what we have discovered about the nature of political power and the structure of power in society. There I will say more about the relationship of individual preferences and interests to one's position in the social structure and how this relates to individual and group power. This will specify some of the limitations of individual and group outcome power. The rest of this chapter will look at previous rational choice attempts to formulate political power, first through the power indices approach which gave birth to formal coalition theory (Riker, 1962; Laver and Schofield, 1990) and secondly through Harsanyi's bargaining approach to political power. Whilst the power index literature has been influential in certain areas of political science, both it and Harsanyi's bargaining approach have been largely ignored by the mainstream power debates. There are two good reasons for this: first, the economists do not explicitly address the concerns of the mainstream debate; second, there are problems with their conceptualization of power. I believe that the mainstream debate should absorb the insights of the economists, note the conceptual problems and overcome them, and identify the limits of a rational choice 'resource' approach. The next two sections will briefly explain and criticize these two approaches. Later chapters will make use of their insights.

4.2 Preferences and power indices

The power index approaches to measuring political power are versions of outcome power because each index attempts to measure the power of individual voters in relation to their preferences for some outcome and the effect they have on attaining that outcome. The power of each individual voter is some function of the number of voters in the electorate and the voting rules under which the outcome is determined. The most discussed index is that of Shapley and Shubik (1969; Shapley, 1967, 1981), which is the subject of my critique, though my criticisms are supposed to apply *mutatis mutandis* to all the indices.[1] Each index does not always calculate the same power to each voter under the same decision rules, a point which has some consequences for constitutional design if equality of voting is one of the requirements (see Grofman and Scarrow, 1979, or Morris, 1987, for discussion). However I am not concerned with their differences here but with what the indices have in common.

The importance of individual preferences to power relations has long been recognized. Jack Nagel (1975, 29) has defined power in terms of a causal relation between the preferences of an actor and some outcome:

> A power relation, actual or potential, is an actual or potential relation between the preferences of an actor regarding an outcome and the outcome itself.

The power he has in mind is 'power over' rather than 'power to', since it is restricted to the state of another social entity: 'the behavior, beliefs, attitudes, or policies of a second actor' (Nagel, 1975, 29). However, the equation of preference and outcome is still made. But this definition is problematic, for it is not restricted by what one's preferences happen to be. If I am unable to do x then I do not have the power to do x whether or not I happen to want to do it. One does not gain or lose power merely by changing one's preference ordering. Thus the potentiality is what one *could* do dependent upon one's preferences. A 'modified Nagel definition' overcomes this problem:

> A power relation is a causal relation between the actual or potential preferences of an actor regarding an outcome and the outcome itself.

The important part of the redefinition is the causal aspect. The preferences of the actor help to cause or could cause the outcome. I will return to this definition later.

The index approach to power suffers from the same problem of preferences. However it is worth considering for it can prove a useful approach to power if we give it a treatment similar to that which we have just accorded Nagel's definition. The power index approach treats power in terms of the rules of voting within committees. The power of any one voter is then determined by the probability that her vote will be decisive in securing some outcome where only voting determines outcomes. Shapley–Shubik imagined a group of individuals all willing to vote for some measure. They vote one by one, and as soon as a bare majority is reached the measure is declared passed and voting then ceases. The last person to vote for the measure is given credit for having passed it. They call this person the *pivotal voter* or *pivot*. In reality, of course, few voting schemes operate in this way. Normally the votes are counted after all have been cast, so that people do not know how others have voted prior to voting themselves. However Shapley–Shubik imagined the voting process in this manner purely to set up a way of looking at the power of each voter, and their index does not rely upon such a voting method. Shapley–Shubik suggest that the power of any individual voter is how often that person is, or could be, the pivotal individual. The pivot can then be defined:

$$P_i = \frac{m(i)}{n!} \tag{4.1}$$

Here P is the power of a voter i in a set of voters $\{1, 2, \dots, n\}$ and $m(i)$ is the number of times that i is *pivotal* in securing that outcome ($n!$ means n factorial and if $n = 4$ then $4! = 4 \times 3 \times 2 \times 1 = 24$). Being pivotal is defined: when the voting rules define q votes as a winning number,

$$\frac{n+1}{2} \leq q \leq n \quad \text{or} \quad \frac{n}{2} + 1 \leq q \leq n \tag{4.2}$$

The pivotal position is the qth position in any ordered sequence of votes, there being $n!$ ordered sequences. Thus:

$$\sum_{i=1}^{n} P_i = 1 \tag{4.3}$$

Under this definition a voter's power is determined by the number of times she is pivotal in relation to the number of possible ordered sequences. In other words the power of any given voter is the probability that that individual is the last member of a minimum winning coalition. The power of members of a committee always sums to 1.

For a committee of three members, a, b, c, the potential ordering of votes is:

$$a\ b\ c$$
$$a\ c\ b$$
$$b\ a\ c$$
$$b\ c\ a$$
$$c\ a\ b$$
$$c\ b\ a$$

In each case the middle voter is pivotal and a, b and c are each the middle voter twice, there are six orderings and thus $i/n! = 2/6 = 1/3$. Thus each voter in a committee of three has Shapley–Shubik power to the amount of 1/3. This seems an immensely trivial result for something I have explained so laboriously. It seems to be less trivial when we consider weighted voting. Imagine a committee of three with weighted voting, $a = 50$, $b = 49$ and $c = 1$ (the example is from Riker, 1969b). The weighting can be explained by imagining the committee is a parliament of 100 members and a, b and c are parties with, respectively, 50, 49 and 1 members of parliament. The set of winning sequences is the same as above, but here a is required in order for a sequence to win for $b + c = 50$ and thus can only form a blocking coalition. Thus b is pivotal once (the first in the order), and c once (the second in the order) and a is pivotal in all the others. Thus the power of $a = 4/6 = 2/3$, $b = 1/6$ and $c = 1/6$. This result seems less trivial, for it shows that, despite its greater strength in terms of votes, b is no stronger than c. This can help explain why small parties in legislatures often have power well beyond their apparent voting strengths. It is from this perception that formal coalition theory has developed (Riker, 1962;

Laver and Schofield, 1990). The criticisms of the Shapley–Shubik index below thus reverberate through coalition theory too.

Perhaps the most damaging criticism of the Shapley–Shubik account of power is that it does not measure power (Barry, 1980). The number of times someone is a pivot in any possible sequence tells us how often that person could wield power if it is known that the person is a pivot in the Shapley–Shubik sense. It is not a measure of their power as such. We can see this if we consider the case of someone who, as it happens, is always a pivot.

The US Supreme Court has nine Justices. Careful studies of voting in the Court have shown that Justices tend to vote on ideological issues as might have been predicted from their previous reputations as liberals or conservatives (Segal and Cover, 1989). Imagine there are two blocs of four. One bloc is made up of four liberal Justices and one bloc is made up of four conservative Justices. There is one maverick Justice (hereafter MJ) who sometimes votes with the conservatives and sometimes with the liberals. On each case before the Court there are only two possible verdicts: either in favour of the Plaintiff or in favour of the Defendant (though each Justice may give different reasons why they voted one way or the other). It might be argued that MJ has more power than the others because she gets what she wants 100 per cent of the time, whereas the others get what they want only (say) 50 per cent of the time each (Barry, 1988, 345). On our revised Nagel definition of power MJ is more powerful than the other Justices, for her actual preferences cause the outcome gained whilst her potential (opposite) preference would also cause her preferred outcome to be gained in each and every case. But this is only so because the preference structure of all the Justices, including MJ, is what it happens to be. Peter Morriss (1987) wants to capture this happenstance within a revised Penrose index (an earlier and, in Morriss's view, superior index to Shapley–Shubik). Morriss (1987, 169) wants to investigate the power of actors

> given the actual (or predicted) voting pattern of all the other voters. The assumption that all configurations count equally has to be replaced with an assessment of which configurations will occur, and how often.

Morriss calls this 'ableness power', after the distinction between 'ability' and 'ableness' discussed above. The test of the ableness power of a maverick voter is whether or not she has power because she is pivotal with a known probability. Does MJ have greater power than the other Justices in this example? Would we think that MJ was more powerful than all the others if she always voted with the conservatives? Then five Justices would get what they wanted 100 per cent of the time and four of them 0 per cent of the time. MJ only appears to have more power because she happens to have been denoted in the example as MJ: that is, she has a different preference structure from the others. Her different preference structure does not, in fact, give her greater power over the results, since the two blocs have their *reasons* for voting together and she has her *reasons* for voting as she does. She only appears more powerful because the preferences of the other two blocs are taken as given but hers is not. When MJ votes with one of the two blocs then any of the other four Justices in that bloc has an equal claim to be pivotal. MJ may be able to claim to be pivotal 100 per cent of the time, the other eight Justices only 50 per cent of the time; but that does not give MJ more power, for the other Justices *could* scupper her plans on each vote. The fact that they never do is simply not relevant to whether or not they *can*; and what they can do is a measure of their power, for power is a dispositional concept. MJ is just lucky to get what she wants all the time – and luck is not the same as power.

MJ can be said to have 'outcome power', for she does have the ability to bring about or help bring about the outcome she wants. But she does not have more of this outcome power than the other eight Justices. For what she brings to each occasion of voting is her one vote, which is the same as what the other eight Justices bring. Outcome power, as argued above, reduces to the resources of the actor and, in these examples, a vote is all the resources each voter has.

The fact that MJ always gets what she wants is not enough to demonstrate that she has power. She only gets what she wants because of the happenstance of the two tied blocs. Whether she can get what she wants depends upon what happens to be the distribution of preferences amongst the Justices. This is a form of luck (see below). It is true that one may become powerful through luck, but that is not the same as being lucky without power. MJ is not lucky that she is powerful and can thus get what she wants; for her luck over the distribution of preferences does not affect the basis of her power, viz. one vote out of nine. Rather she is lucky that she happens to get what she wants despite having no more power than the other Justices, for she, by hypothesis, has no power over the distribution of preferences and if this should happen to change she would no longer get what she wants. MJ may have been lucky to have been made a Justice in the first instance: the basis of her one vote would then be luck, but that is a different matter entirely.

Happening to be pivotal or being lucky that one is pivotal brings power, but these circumstances are not in themselves a form of power; rather it is the *realization* that one is pivotal which brings power. In other words, being pivotal is a *resource* which may be used by the holder to a greater or lesser extent. If MJ votes entirely on her preferences then she has no greater claim to be pivotal than the others, for she is behaving just as they are. Any one of the five winners can claim to be pivotal (that is, none is), for they are indistinguishable except by definition. The greater power of pivotal voters, for example small parties in hung parliaments, comes about not because they happen to vote one way or another but because their actions change because of the realization that they are pivotal. Pivotal voters behave differently from non-pivotal ones. They can use the fact of their critical position in order to achieve a greater number of outcomes than they otherwise could. (Not necessarily, note, a greater number than non-pivotal voters – large parties still write most of the legislation emanating from hung parliaments – but more than the pivotal voter could have got if she were non-pivotal with the same amount of luck.)

This greater power is power to bring about outcomes beyond that of the particular issue under vote. Consider the case where the two blocs of four voters are parties with party discipline and the extra voter is an 'independent'. Here the independent does have power to get some of her desires, for she can use her position as an independent to bargain with each of the two blocs. She may vote with one bloc on one issue over which she is indifferent (or leads the blocs to believe she is indifferent) in return for their support on issues where she holds stronger preferences than they. She thus has power to achieve some of the outcomes she wants by agreeing to vote one way or another. But here we have power as ability – the power to bargain – and not ableness – power dependent upon the way people happen to vote. For the independent voter will only choose to support one bloc or the other on each vote according to (a) her preferences with regard to the issue in question, and (b) whether she can persuade either bloc to agree to vote for some other policy which she wants in return for her support on this issue. The measure of her power is how much she can get of what she wants through her ability to bargain with the two blocs. If she can get everything she wants simply by voting with either of the blocs on each issue, then she is just exceptionally lucky; but her *power* is her capacity to get new issues onto the agenda and to push through those policies she supports. She may vote with one bloc on issues to which she is indifferent or even opposed, as long as the strength of her dislike of those policies is less than the strength of her desire for the issues she is able to get onto the agenda. Formal coalition theory may thus give important insights into the behaviour of leading actors and their powers (McLean, 1987, 107–20).

The recognition that one is pivotal produces the ability to get other things done. It is a *resource* which brings power. One may become pivotal because one is the last person to be asked to vote a particular *way* and realizes how important one's own vote is. In some Indian states villages vie with each other to be last to promise their votes, for that way they can gain other (usually monetary) rewards (Wade, 1987). In doing so they recognize that being last they may be pivotal and being so may bring further rewards. Bargaining strategies, such as pre-commitment or keeping one's true preferences hidden, are what gives the 'pivotal' voters power. The problem with power indices is that they do not denote the pivotal voter by anything other than definition. The pivotal voter is always the one with the floating preferences; but power is the power to get what you want, not getting what you can in order to prove you are powerful (Barry, 1980). This flaw in the power index approaches is one which haunts much of mathematical economics and political science.

4.3 Distinguishing luck and power

Brian Barry's important (1980) critique of the power index approaches argues that they concentrate too much upon outcomes and not upon power as such. In short, Barry argues that one does not need to have power in order to get the outcomes one wants. Rather one may get those outcomes just through luck. This is what MJ relies upon. Barry (1980, 183) asks:

> Is it better to be powerful oneself or to have powerful friends? Must the power of a group be conceptualized as the sum of the power of each of the individual members of the group, or could a group be powerful whilst each of its members is individually powerless?

Barry's questions are the key ones which this book attempts to answer. Part of my contention (and in a less-developed form of Barry's articles) is that much of the debate between pluralists and elitists, marxists and statists, arises because they are often arguing at cross-purposes. Outcome power is confused with social power; luck confused with power of both sorts; and the power of a group confused with the power of the individuals it contains. I will try to disentangle some of these confusions here. More detailed discussion of the mainstream protagonists comes later.

If the conservative bloc on the Supreme Court have a clear majority, say six to three, then as a group the conservatives are clearly powerful. On all 'ideological' decisions they can get what they want. However each conservative Justice, as an individual, has exactly the same power as every other, and as each of the liberal Justices: viz. one vote each. Or rather, more carefully, we can say that each Justice has the same formal voting resource as every other: viz. one vote each.

Barry may find the careful conclusion acceptable. He is unlikely to accept the previous sentence. For Barry believes that political power must not be defined in terms of getting outcomes, but getting outcomes despite resistance. He defines power (1980, 185):

> political power…[is]…the ability of an individual or of a group to change the outcomes of some decision-making process from what they would otherwise have been in the direction desired by the person or group, where the decisions made are binding on some collectivity.[2]

This is a basically Weberian definition of power. It is both more general and slightly weaker, for Weber (1978, 53) says that power is

> the probability that one actor within a social relationship will be in a position to carry out his own will despite resistance, regardless of the basis on which this probability rests.

Powerful people and groups may not be able to carry out their precise will in any given situation, but they may, as Barry suggests, be able to move the decision in the general direction of their will. We have seen that just getting what you want is not enough to demonstrate power. One can achieve one's desires with no power at all if one is exceedingly lucky. Barry defines luck as the probability of getting what you want without trying. Success is how often one gets what one wants if one tries. The difference between success and luck is an individual's decisiveness. So success = luck + decisiveness. Thus the notion that Shapley–Shubik were trying to develop with the number of sequences in which some voter is pivotal divided by the number of possible sequences is, in Barry's terms, a measure of their decisiveness. And again like the Shapley–Shubik power index, for any given individual each of these three measures will take a value between 0 and 1, but the scores of all members do not have to sum to 1.

The luck of which Barry writes must be carefully distinguished from another sort of luck discussed by egalitarians (for example, Roemer, 1986a; Cohen, 1989): the luck of being the particular person one happens to be. We may call this luck 'Personal Identity Luck'. Barry Luck is getting what you want without trying, no matter who you are; but Personal Identity Luck is tied to the internal and, to some extent, the external resources that one has through being the person that one is. Thus one may be lucky that one is beautiful, talented and rich, or unlucky to be ugly, untalented and poor. This type of luck is summed up in Woody Allen's joke that his one regret in life is that he is not someone else. In Chapter 7, I will show that the gap between these two sorts of luck is not as great as it might at first appear because of the existence of 'systematic luck' introduced in Chapter 6. This suggests that the relationship between Barry Luck and Personal Identity Luck is not entirely contingent when the former notion is discussed in relation to individuals as denoted by their institutional and social location.

Brian Barry writes (1980, 348) that power is not like luck, success or decisiveness in that it is a capability and not a probability:

> An actor's power is his ability to overcome resistance – not his probability of overcoming resistance. For the probability of overcoming resistance depends on the probability of encountering resistance, and not just any resistance but resistance of the right amount.

This is true enough. An actor's social power depends upon his resources which are the basis of his capabilities. But Barry is happy to define luck, success and decisiveness as probabilities, although they are not ordinarily thought of as such. One's success is usually thought of as getting the outcomes one wants; rather than, as Barry expresses it, the probability of getting the outcomes one wants.[3] Similarly, Barry defines luck as the probability of getting what you want without trying, whereas ordinarily luck is thought of as getting what you want without trying in situations where the probability of getting what you want is low. He defines decisiveness as the increase in the probability of getting what you want if you try, whereas in this context it would be thought of as the characteristics of the person who tips the balance.

The fact that Barry has not defined these terms in the way we ordinarily think of them does not, of course, totally vitiate his argument. His definitions can be thought of as technical definitions, but his approach does raise the question of whether we can handle power in this way. In fact we can do so. Whilst recognizing that power is a capability we may still define it as

> the probability of getting what you want if you act in all possible worlds which are the same as the actual one with the exception of the preferences of other actors.

This definition retains the capability aspect of power whilst building in a probabilistic analysis of an actor's success as a result of his capacities.

It retains the capability aspect because it recognizes outcomes which you want as being a result of your actions rather than through luck. It does not include his luck in the actual world, for the definition covers success in all possible worlds, thus including those worlds in which he is not lucky enough to get the outcomes he wants. Thus this definition recognizes that, whilst I may be lucky in this world and get what I want without having to act, I am still powerful because, if not lucky, I would have achieved my aims through my actions. It is of course possible to be powerful *as well* as lucky.

This definition may now be compared with the revised Nagel definition:

> A power relation is a causal relation between the actual or potential preferences of an actor regarding an outcome and the outcome itself.

Though very similar, there is an important difference between them. The revised Nagel definition holds the world the same but varies the preferences of the actor under consideration. The revised Barry definition holds the preferences of the actor the same and varies the world with regard to the preferences of other actors. Put into the form of a question, Nagel revised asks, 'Given what everyone else wants, what can I get done?'; Barry revised asks, 'Given what I want, can I get it done no matter what others want?'. Either is a fair enough question to ask and either way of looking at power is acceptable. The difference is that Barry's explicitly brings conflict into the definition, Nagel's does not.[4] In terms of the definitions discussed above, the revised Nagel is 'power to' or 'outcome power'; Barry's is 'power over' or 'social power'.

The decisiveness and luck of an actor varies according to the preferences of other actors, but an actor's power remains the same. It is a disposition, analysable counteractually by taking into account possible preference changes.[5] This formulation deals directly with outcomes. Being powerful is getting what you want, but possible resistance is also built into the definition. This analysis, of course, leads us to a trivial truth, though at first it may appear to be a trivial falsity. Under the revised Barry definition no one has power, for no one can get anything done in every possible world: thus no one has any social power.[6] However, this trivial truth leads us to an important one. Every powerful person is powerful because of the resources they bring to a bargain with other actors. Social power always depends upon a coalition of mutual or allied interests. Consider a dictator who gets every outcome he desires. How does he manage it? He relies upon so many other people: his army, his police, his secret police, his cabinet, and so on. All these other people, or some subset of them, could conspire and overthrow him. However dictators survive by forming coalitions, in the technical sense of the term, with these other people and by stopping rival coalitions forming by sowing doubts in others' minds and turning potential partners against each other. The dictator offers negative and positive incentives to all the other actors in order to gain their support and stop their challenge. In order to understand even the most obvious examples of social power we need to understand the nature of coalition formation and the nature of bargaining. All political power is a form of reciprocal or bargaining power.

One may sooner be lucky than decisive, for then one gets what one wants without trying. One may sooner share the same interests as the powerful than actually be one of the powerful. Thus individuals must make judgements about getting what they want by taking into account the interests they share with others, and judgements about power must be judgements about groups rather than individuals. What individuals take into account in these calculations are the resources that others could bring to bear against them in any social situation. The power of others is assessed solely in terms of their resources. These resources include both 'external' resources – money, legal or institutional position, for instance – and 'internal' resources such as physical strength, determination or persuasiveness. My account of power entails considering the resources enjoyed by different individuals and groups in society. Most important are their institutional resources, a vague phrase here but one to be cashed later when I shall also deal with the charge of the 'vehicle fallacy' – the equating of resources with power.

We can see that the power index approach places too much stress upon outcomes in its attempt to measure individual power. Getting what you want is a function of both power and luck. Morriss's attempt to assess the likelihood of success by incorporating the preferences of actors (foreshadowed in a different way by Axelrod, 1970) takes the approach further down the wrong line. Rather we should see one's voting strength as one resource and one's 'pivotality' as one's luck, though a form of luck which may be utilized to one's advantage together with all the resources one has. J.C. Harsanyi generalizes the Shapley–Shubik indices approach and attempts to formalize the resources of actors in order to study their power within a bargaining model.

4.4 Reciprocal or bargaining power

In the next two sections I will consider the elements of a bargaining model of social power based upon the work of Harsanyi. This will be done in a non-technical manner, for the importance of the bargaining models for empirical political scientists lies not in the precise formulation of the equations but rather in their individual elements. There are far too many practical and technical problems involved in quantifying power relations by placing actual values upon the variables in the equations for those equations to be useful for actually measuring power. The value of the formal models derives from their careful conceptualization of power which allows us clearly to recognize the factors which need to be taken into account. Formalization is an important conceptual tool. After discussing the models I will turn in the final section to the way in which we should approach power studies in society. Chapters 5 and 6 will examine some of the problems of previous power studies, largely the result of poor conceptualization, and attempt to draw some conclusions from the power debate.

The social power of one actor A over another actor B is the ability of A deliberately to alter the incentive structure of B in order to get B to do something that she would not otherwise do. There are four main ways in which A can deliberately alter B's incentive structure. First, A may persuade B that an option in B's preference schedule is not what it seems, and should be raised or lowered in her estimation. Second, A makes an offer to B to raise some option in B's estimation. Third, A may make some threat to B, thereby lowering some option in B's estimation. Fourth, A may make both a threat and an offer – sometimes called a 'throffer' (Steiner, 1975) – both raising and lowering options in B's estimation. The first of these four ways of affecting B's incentive structure does not have to be explicit or recognized by B. For example, A may persuade B to alter the order of options by changing the information available to B without B's cognizance. However, B's preference ordering is changed these possibilities may be represented in the following *manner*.

If B's original preference schedule is

$$\{\ldots a > b > c > \ldots\} \qquad (4.4)$$

then A persuades B if he brings about this reversal in B's preference ordering:

$$\{\ldots b > a > c \ldots\} \qquad (4.5)$$

Some action on the part of A causes this change in B's preference ordering. A may do this by pointing out bad attributes in a (turning it into a'), good attributes in b (turning it into b'), or both:

$$\{\ldots b > a' > c \ldots\} \qquad (4.6)$$

$$\{\dots b' > a > c \dots\} \tag{4.7}$$

$$\{\dots b' > a' > c \dots\} \tag{4.8}$$

These three ((4.6)–(4.8)) represent one way in which we may be persuaded to change the order of options. In these examples the order of the options has been altered because we have been persuaded that there are descriptions of those options of which we were unaware. For example, I might prefer to buy the teddy bear on the right, for it is cheaper than the one on the left; but then be persuaded that I prefer the one on the left, for it has no dangerous pins holding the eyes which could injure my child. Or I might be persuaded not that b is preferable to a in itself, but that b has consequences of which we are unaware. For example, I might be persuaded that, if all the employees in a factory take a 10 per cent pay rise, consequently the company will lose money, go out of business, and I will lose my job. Persuasions of this type are like warnings and have three forms: a negative form, a positive form and a combination of the two.

$$\{\dots b > a + x > c \dots\} \tag{4.9}$$

$$\{\dots b + z > a > c \dots\} \tag{4.10}$$

$$\{\dots b + z > a + x > c \dots\} \tag{4.11}$$

Here A has pointed out to B that option a will (probably) lead to the undesired consequence x; option b will (probably) lead to the desired consequence z; or both.

There is an important difference between (4.6)–(4.8) and (4.9)–(4.11). In (4.6)–(4.8) the preference ordering was fundamentally altered as the positions of a and b were swapped over. However in (4.9)–(4.11) the ordering was altered only to the extent that new consequences were added. The original ordering still remains, for $\{\dots a > b > c \dots\}$ is buried in $\{\dots b + z > a > b > a + x > c \dots\}$. Thus the first sort of persuasion differs from the latter in that in the first B has changed her belief about a or about b or about both, whereas in the latter she has merely become aware of unrecognized consequences. (Where those unrecognized consequences are strictly necessary features of a or b, then (4.9)–(4.11) collapse into (4.6)–(4.8).) The latter forms of persuasion look very much like threats, offers and throffers.

Threat: $$\{\dots a > b > a + x > c \dots\} \tag{4.12}$$

Offer: $$\{\dots b + z > a > b > c \dots\} \tag{4.13}$$

Throffer: $$\{\dots b + z > a > b > a + x > c \dots\} \tag{4.14}$$

These representations of threats, offers and throffers do not differ in any way from the second sort of persuasion. Rather the difference is the nature of the act that A performs in order to bring about these changes. Under persuasion as I have defined it A is merely making a prediction. Under threats, offers and throffers A is making a conditional promise. A is saying that, if B attempts to bring about a, A will arrange it that x follows (threat). Or if B brings about b, A will arrange that z follows (offer). Or both (throffer).

It is important to note that there are potentially two acts of A involved in both threats and offers. There is action which is a threat/offer and action which implements a threat/offer. The power of A over B depends upon the success that A has in getting B to do what A wants her to do. Successful threats must, at least, place $a + x$ below b; successful offers must, at least, place $b + z$ above a. In order to achieve this result threats and offers must be credible: B must believe that A will carry out the threat x, and provide what is offered by z. Thus the power of A depends both upon his resources and upon what B knows of those resources. I will explain how we measure individual power by considering a simplified version of Harsanyi's bargaining theory and demonstrate its limitations.

Harsanyi (1969a, 229–30) identifies four main types of influence techniques available to the power-holder, A.

1. A may supply information or misinformation to B in order to affect B's calculations of the opportunity costs of action. A's costs are collecting and communicating this information to B. Information is a very important part of power relations. Collecting information generally holds costs for individuals and rational actors may decide that the costs of collecting information outweigh the benefits it may provide. Goodin (1980) describes various ways in which hiding, misinforming or telling half-truths may be used to manipulate others and he gives many examples where various state actors have used their near monopoly of information to their own advantage. Generally speaking, governments and more particularly bureaucrats have privileged access to information and may thereby create costs too great for ordinary citizens to contemplate to discover the truth. One of the important ways in which pressure groups gain power is to break down the informational barriers put up by government. Also important is not only a free but also an investigative press. For investigative journalists collecting information is not a cost (at least not in the same way) but a benefit, for that is what they are paid to do.

2. A may have legitimate authority over B. Gaining this legitimacy may involve costs to A, though those costs are ones which are unlikely

to enter as costs into A's calculation at this point.[7] The legitimacy which all state actors enjoy over other actors in society is an important resource which is utilized to the full. A major criticism of those pluralist models which see the government or state as a neutral arbiter between competing groups is that it does not take account either of state actors' own interests or of the greater power these actors have because of their legitimacy. Of course, other groups in society may gain legitimacy too. Neo-corporatists have distinguished organized groups which enjoy corporatist relations with the state from organized pluralist groups which have a different relationship. One of the features of the former is the different legal status they enjoy. Close institutional links and legally defined authority and responsibilities provide such organizations with greater resources to press demands. Their role in policy implementation as well as policy formation gives power which the excluded groups do not enjoy.

3. A may provide B with a number of unconditional incentives (negative or positive) which affect B's calculations of the opportunity costs of his action. They are unconditional in the sense that B has to bear the costs, or receive the advantages of them whether or not he does as A wishes. For example, A may provide certain facilities for B which make it easier or cheaper for B to act in ways in which A desires. By removing or adding facilities A may also make it more likely that B will be forced to act in ways which A desires. Either way, A has to bear the costs of attempting to influence B. For example, by encouraging home ownership the British Conservative government hoped to change the attitude of former council-house dwellers towards rent subsidies in the public sector. Similarly increasing share ownership changes attitudes towards taxation policy. Trade unions may affect the attitudes of their members, or managers their workers, by offering different sorts of incentives for actions one way or another. Thus the incentive structure of individuals are altered by the actions of the powerful. In all of these cases the opportunity costs of acting are altered.

4. A may supply B with conditional incentives (positive or negative). Here A either promises to reward B if B behaves in a certain way, or promises to punish B if he does not behave in a certain way. A's costs will vary depending upon the way in which B behaves. This is a most obvious and blatant use of power, and the sort of power use which is familiar to everyone – and incidentally the only type of power usually associated with individualist methods. In the 'actual expenditure approach' (Baldwin, 1989, 86) we calculate the costs which A actually incurs as a result of B's actions. Here successful threats and unsuccessful offers are virtually costless.

There are two problems with the actual expenditure approach (Baldwin, 1989, 86–7). First, it fails to consider the reduction in A's resources at the moment of making the conditional commitment. If someone has committed resources to some future contingency then they have fewer resources to commit to other things. One cannot make too many threats or offers at the same time. Second, it fails to distinguish between lucky people and powerful ones.

A better way is to calculate costs by the expected costs of the attempt (Harsanyi, 1969a, 227). These costs are valued by the risks that A runs when making the threat or offer and not the ones he may actually have to pay out. Thus the costs to A must take into account:

(a) the costs of communicating the commitment,
(b) the costs of making the commitment credible,
(c) the costs of monitoring B's activities, and
(d) the costs of carrying out the commitment.

(a)–(c) are borne by A regardless of whether B complies, but (d) is dependent upon B's response, and must be weighted by the probability that A will incur these costs. Three probabilities are relevant: (i) the probability that B will comply, (ii) the probability that A will carry out the commitment, and (iii) the probability that A will carry out the commitment regardless of B's actions.

There is an important asymmetry between threats and offers because offers cost more when they succeed and threats cost more when they fail. Thus increasing an offer is likely to increase the probability of incurring the costs of success: whilst increasing a threat, as long as it remains credible, is likely to decrease the probability of incurring the costs of carrying it out. Their credibility is the only limitation on threats. The cost of armed rebellion by the 'Keep Sunday Special' campaign, should the 1986 Shops Bill have been passed, is so great that such a threat would have been incredible. Schelling (1966, 35–6) argues that some threats are inherently incredible, others inherently credible. If there were not such limitations upon the credibility of threats then it would always be worth one's whilst to make the threat as great as possible. Hobbes's state of nature is so nasty because there is no limitation on the credibility of threats therein – you may as well kill someone as refuse to speak to them again – which is one of the problems of Hobbes's argument.

One way of making a threat credible is to precommit oneself to a course of action which necessitates carrying out the threat should the given circumstances arise. Governments may precommit the state in this way through legislation to ensure that breaking the law entails heavy punishment. But if the penalty seems out of proportion to the crime, juries may refuse to convict and judges may try to avoid meting out such a harsh punishment. It is difficult to precommit others on one's own behalf. Any attempt to precommit oneself is merely another form of threat: for example, trade unions which, prior to negotiations with management, vote on strike action to be taken if certain conditions cannot be obtained for their members.

There are several important behavioural consequences of the asymmetry. Will a punishment and a reward of the same value to B be equally effective at getting B to do something he would not otherwise do? We could make this so by definition, by defining the value to B of rewards/punishments in terms of their behavioural consequences. This definitional move is tempting, but it does not allow for the possibility of stubborn people who will not do something because they have been threatened with punishment if they do not. Rational stubbornness is an important consideration which places limitations upon the bargaining approach to power. However, from A's point of view, this question is not the important one. A does not care how much B values promised rewards or punishments, but only about B's responses to them. What matters for our calculation is not B's valuation of the punishment or reward, but the costs to A of actually carrying them out. If we assume for a moment that B is equally likely to carry out A's preferred course of action whether the incentive A provides is in the form of a threat or an offer, we can make some calculations about which form it will take. The more likely it is that B will comply with the incentive, the more likely it is that it will take the form of a threat. The more likely that B will not comply with the incentive, the more likely it is to be an offer. For if A's threat is successful, she does not have to bear the costs of carrying out the punishment; whereas if A's offer is successful, she must provide the promised reward. (I say 'have to' here because I am assuming that if A does not carry out her threats and offers they will not remain credible.) Thus it is in B's interests to indicate that he will not comply with proffered incentives, for then the incentive is more likely to be made in the form of an offer. But consequently, it is also in A's interests to threaten costlier sanctions than she would want to offer. However, the costs to A in carrying out threats or providing rewards are unlikely to be symmetrical with B's gains and losses with regard to those same punishments and rewards.

All that we can conclude is that it is in B's interest to promote the idea that he is more likely to do what A wishes if he is offered a reward than if he is threatened. And A will not make offers which are worth as much to B as the utility to be lost when she threatens.

Harsanyi suggests that the amount of an individual's power is the difference between two probabilities: the opportunity costs of attempting to influence someone, and the opportunity costs of refusing to comply. He says:

> As these opportunity costs measure the strength of B's incentives for yielding to A's influence, we shall call them the *strength* of A's power over B. (Harsanyi, 1969a, 227)

and

> [M]ore precisely the *costs* of A's power over B will be defined as *the expected value* (actuarial value) of the costs of his attempt to influence B. It will be a weighted average of the net total costs that A would incur if his attempt were successful (e.g. the costs of rewarding B), and of the net total costs that A would incur if his attempt were unsuccessful (e.g. the costs of punishing B). (Harsanyi, 1969a, 227)

Harsanyi is aware that all power relations are bilateral. Each player may threaten the other if only in the sense of refusing to cooperate in some way. It is hard to disentangle the power of each from the power of one over the other (Simon, 1969). We may see how Harsanyi's bargaining model works in a simplified form with regard to threats only. We can then demonstrate its limitations, which stem from the ability of each to impose costs upon the other by refusing to comply.

If we wish to measure the extent of A's power over B, we need to assess the value to B of A's threats against him. If B acts to maximize his expected utility then the action he would choose in the absence of A's threat will have a *higher* utility than the action into which he is coerced. If t_1 is the measure of the disutility to B of the sanction A threatens, then the threat will be successful if $t_1 > u_1 - u_2$, where u_1 is the utility to B of carrying out his preferred action without the threat, and u_2 of carrying out the coerced action. The difference $u_1 - u_2$ that A can make to B's welfare is the measure of A's power.

Modern non-cooperative bargaining theory (for example, Rubinstein, 1982; Sutton, 1986) shows that A does not necessarily have the amount of power suggested by this account. B may rationally refuse to comply with A's threats despite $t_1 > u_1 - u_2$. Assuming that there are costs to A of carrying out her threats, there is no advantage in threatening those who will not comply. If there is a potential pool of threatened persons then A would sooner threaten those who will comply. There is thus an incentive for B to build up a reputation for 'stubbornness' where 'stubbornness' is defined as: B is being stubborn when he acts so as to bring about y rather than x, when B prefers y to x but knows that y will cause sanction z and $x > y + z$.

In order to understand stubbornness we have to understand rational stubbornness; otherwise we could not demarcate cases of stubbornness from conditions of $y + z > x$. Rational stubbornness occurs when an individual can see potential long-term benefits in acting stubbornly. Individuals recognize that a reputation for stubbornness will mean that they are threatened less often and thus may carry on doing what they prefer. If a strong A realizes that a weak B is unlikely to respond to threats then A is more likely to make an offer to B in order to get what A wants. This may happen even if there is not a pool of potential victims. B might still be stubborn in order to try to make A realize that there is no point in threatening him. However a credible threat to kill B could not be rationally resisted if B values his life above the action A is trying to make him perform, for there can be no future benefits to B in so resisting.

The importance of reputation in n-person game theory is now widely recognized (for example, Roberts, 1985) and does place limits upon any programme of measuring power by looking at the resources of the players. We might say that if all players had complete and perfect information then reputation would no longer be important, for people would no longer have any scope for building up a reputation as anything other than what they are. Thus the actual importance of reputation in game theory arises from asymmetries in players' knowledge. However complete information requires all the following statements to be true: 'You know who I am', 'I know you know who I am', 'You know I know you know who I am', *ad infinitum*. This is a very strong assumption so we might just as well say that the importance of reputation depends upon the actions of individuals as much as upon their resources. I find the limitation comforting for two reasons. First, it ends determinism in the models (something which seems inherent in Harsanyi's actual programme). A player can (rationally) decide to be stubborn at any point in an iterated game, and this therefore places limitations upon what we may predict about actions even given full information about the 'external' resources (Roemer, 1986a) of all players. Second, it explains, within the terms of the rational choice approach, an often used argument against Harsanyi's reasoning on power ascription. Some political scientists have claimed that people just do not behave in the way in which Harsanyi's resource approach predicts. Therefore individual 'power' cannot be measured by the 'external' resources that individuals bring to bear in any situation of conflict. This is sometimes called the 'vehicle fallacy', for it is supposed to equate power with its vehicle (Morriss, 1987, 18). The fact that some people have certain resources and others fewer does not demonstrate the power of one over the other. The greater resources of the first group may be the means by which that group is able to have power over the second, but it is not the same as that power. The fact that people are sometimes stubborn may be said to prove this. The fact that other 'internal' resources may be brought to bear (personality, degree of risk-aversity, stubbornness, and so on), which can only be imperfectly predicted by past behaviour, does not vitiate the Harsanyi approach *in toto*; it only places limitations upon it, and, as I have stated, I see those limitations as an advantage. Nevertheless, the best starting-point for study of the relative power of groups in society is an examination of their resources, given the rules of the game they are playing.

The stubbornness objection does not only apply to threats: people may be made offers 'they can't refuse' yet stubbornly refuse them; every person may have their price but the price for the stubborn may be ridiculously high. The objection also may apply to (1)–(3) as well as to (4), though not quite in the same way or to the same degree. We can see that it may apply straightforwardly to (2) where an actor, B, stubbornly refuses to take notice of the authority that even B acknowledges A has over him. However, generally speaking, the admission of A's authority makes it seem less likely that B will be stubborn, for acknowledging authority is a behavioural phenomenon. We may also see how B may be stubborn with regard to (3). A may try to shift the costs to B of various courses of action. If B is aware that this is what A is doing, then B's stubbornness is explicable in the same way as it is in cases of threats or offers under (4). However, if A's manipulation of B's costs and benefits is done so subtly that B does not realize what is going on, then stubbornness is harder to identify and explain. Here B seems stubborn in his one-minded pursuit of a course of action which appears, given what is known about his preferences, against those preferences. In such a case we either have to fall back on an assumption of preference change, or we identify and explain the putative stubbornness by saying that B has identified the outcome he wants and will pursue that outcome no matter how the world changes in the meantime. In the case of (1) it is harder still to identify stubbornness. Again if B can see that A is trying to hide something or to mislead him then his stubbornness in pursuit of the information, way beyond its expected benefits to him, may be described as stubborn. Of course, if B thinks that A is deliberately hiding something from him, then he may believe that the benefits of knowing may be greater than they are, precisely because A is hiding that information (Goodin, 1980). However, if A manages to increase the costs of collecting information without

B realizing that this is what she is doing, then stubbornness is hard to identify, for it is difficult to see what B is stubbornly doing. If B does not realize that there is information to be collected on some subject, then pursuing that information seems perverse rather than stubborn.

The stubbornness limitation upon the resource approach might appear to make that approach vacuous, since it seems probable that one can rationalize any concept of bargaining power with some sort of uncertainty. As Davidson (1982, 303) says in another context, one may always explain apparent irrationality by moving the boundaries of the rational.

> The underlying paradox of irrationality, from which no theory can entirely escape, is this: if we explain it too well, we turn it into a concealed form of rationality; whilst if we assign incoherence too glibly, we merely compromise our ability to diagnose irrationality by withdrawing the background rationality needed to justify diagnosis at all.

We might get around this problem with broader concepts of rationality than mere connectedness and transitivity, but the danger of vacuity does not trouble me overmuch in an account of the general structure of power in society. It would trouble me in an account of a particular non-cooperative bargain which explained the outcome merely by touting some uncertainty on the part of one or more of the players. In such a case the assumption of uncertainty may save *any* particular explanation of that bargain. However in explaining the more general cases we are not trying to provide a single explanation of unique phenomena. Rather we are trying to provide the outlines of an explanation of all cases. We are looking for patterns and statistical tendencies. If we find them we have found the structure of those general cases into which particular causal explanations may fit. Any seeming vacuity at the general level may then be filled in by the empirical study at the particular level. If, for example, we state that outcome x can only occur if a certain set of players lacks relevant information, behave stubbornly or have intransitive preference orderings, then any actual historical process which leads to outcome x may be studied to discover the form in which one of these has occurred. The general explanation is not vacuous for it gives the framework for the empirical study which will fill in the gaps. However a particular study with such important gaps would be vacuous. For example, if we only explain the behaviour of the National Union of Mineworkers on some issue by saying it lacked relevant information, or Arthur Scargill was stubborn, or the executive had intransitive preference orderings, our individual study would be rather pathetic. Empirical studies are supposed to answer these either/or enquiries; general models may leave them open. The nature of explanation at the two levels does not have quite the same logical form. Hence the simple account of rationality is good enough for the general model.[8]

It cannot be emphasized too strongly that any bargaining model of power involves reciprocity. But this is involved in all attempts to explain the political power of actors. One of the early objections to Dahl's (1968, 1969b) formulation of political power – 'A has power to the extent that she can change the behaviour of B' – was that if A acts to change B's behaviour then B must have power over A too. For if A has to act to change B's behaviour, then by the definition B has changed A's behaviour and hence is powerful. But once we understand the reciprocity involved in all power relations between actors we can see that this is not an objection. All actors involved in relations have some power to help change outcomes and some power over each other, but we are interested in the scope and degree of their power relative to each other. It is surprising how much of the power debate founders on this simple truth, made obvious by the bargaining model which was first published in 1962.

The ability to shift the incentive structure of others is thus an important power resource. Governments, companies, trade unions and other organized groupings have these resources to a greater or lesser extent. An organized group has these resources if it can impose costs upon state actors. Dahl (1961c, 1986) argues that, even where groups do not achieve the implementation of the policies they desire, they still have some power if they can impose costs upon state actors. They may do this by embarrassing the government by disseminating information contrary to that propagated by government, or imposing costs upon bureaucrats through their lobbying efforts. However, the ability of one group to affect another is complicated by the fact that groups are coalitions of individuals. Thus the power of a group is affected by the nature of the coalition – not only the resources that group members bring to it, but also the nature of their common and contrary interests. Harsanyi tries to model this with an *n*-person bargaining model.

4.5 Luck and group power again

Harsanyi's first article on power concerns bilateral bargaining, but bargaining among more than two people is qualitatively different, for it opens up the possibility of coalition formation. Here individual common interests as well as contrary ones come into play. Indeed, whilst stubbornness (or more generally 'free will') ensures that determinate solutions in two-person bargaining cannot be assured, stubbornness is more likely to emerge in n-person cases. Indeed I presented it in an n-person form. Even without the indeterminacy provided by stubbornness, n-person bargaining theory provides indeterminacy, for it may be impossible for stable coalitions to form as the players outbid each other to make alliances in ever-continuing cycles. Harsanyi's second article on power (1969b) considers the n-person case and tries to answer Barry's second question:

> Must the power of a group be conceptualized as the sum of the power of each of the individual members of the group, or could a group be powerful whilst each of its members is individually powerless?

Harsanyi develops a model for explaining the group power of a coalition when no individual members are decisive. He is forced to modify his account (1969b, 239) of the amount of power possessed by each person within an n-person bargaining game because of its qualitative difference. He changes A's power over B with respect to some action x from the increase in the probability that B will perform x to the probability that A's preferred course of action x will be performed owing to the joint action of A and B to bring it about. If there are only two people there is no difference in the two cases (thus the new formulation is preferable), but where more people are involved the new definition allows all their actions to be taken into account. Harsanyi develops a model of bargaining to measure the power of groups of people in coalitions. He wants to measure the power of the coalition to get its 'preferred' policy adopted against opposition from rival coalitions, and the power of each individual within the coalition to affect the form of the 'preferred' policy. The model thus tries to combine the power of a group in relation to other groups with the power of the individuals within a group in relation to each other. Of course, individuals within different coalitions may bargain across coalition lines, which is what makes solutions unstable. One coalition may try to entice an individual from another if this will strengthen the former coalition. That individual will be enticed only if she can get that coalition to adopt a policy closer to the one she prefers than the policy of the coalition she leaves.

We can see that this model is an attempt to generalize the Shapley–Shubik power index to take into account a more subtle view of preferences and one which can encompass resources other than the mere value of an individual's vote. Within this model we have a measure of both an *individual's* power to get her preferences incorporated in the policy adopted by the group and the *group*'s power to get the policy adopted by society. Some individuals have much greater power within a group than others. Their power will depend upon how they can affect other group members' incentive structure. Their power will be based upon the four elements considered in Section 4.4. The most powerful will not only be those with the greatest number of resources but those willing to invest those resources in group aims. There may be several reasons why individuals are not prepared to invest those resources. The first part of their decision concerns the group policy. First, individuals may not contribute because they are reasonably satisfied with the group policy and have no incentive to change it. They are lucky in that regard. Second, even if they are not satisfied with the group policy they may decide that their chances of getting the policy altered are minimal. The second part of their decision concerns the probability of the group succeeding. Individuals may decide that this will not be significantly altered by their help. All of these decisions are concerned with the 'freerider' or 'collective action problem' which is the major block upon group power in society. Group organizations usually end up devoting a substantial proportion of their resources to trying to overcome these collective action problems (Olson, 1971; Moe, 1980). Paradoxically, the happier a member is with group policy the less likely they will participate, the less satisfied the more likely they will take part, assuming that the costs of participation remain stable.

There is no need to go into the full complexity of Harsanyi's model, for he is interested in providing an equation for the amount of power each individual has over any social outcome. For most individuals that power is minuscule. Further, as stated above, we cannot hope to place actually quantifiable values on the variables in Harsanyi's model. Rather the model's utility comes from the clear thinking it engenders by its conceptualization of the elements which must be taken into consideration when studying the power of groups in society. We can adopt a twofold strategy: first, to determine the power of each group member to get a policy adopted that is close to their preferred alternative; second, to examine the power of a group to get that policy adopted by society. In either case we look at the resources that may be brought to bear, in the first place by the individual, in the second by the group. The latter is made up not only of the individual resources that each group member is willing to commit but also of the independent resources of the organization, including the status of the group within the state and society and the quality of its personnel. Barry's question may be answered. Group power does not need to be conceptualized as the sum of the power of its individual members. Generally speaking, in social science, we are more concerned with group than individual power. A group, as well as the individuals within it, may be lucky when the group does not need to act in order to get the outcomes it desires.

4.6 Studying power in society

We should now be in a position to examine some of the issues which separate pluralist from elitist from statist writers. We cannot map the power structure by merely studying who gets what and when. We need to understand why. Pluralists have tended to concentrate upon decisions which affect those groups whose organizations have explicitly lobbied government. However other groups – particularly state actors – may also have benefited. A group which lobbied government may only have succeeded because it was pushing at an open door. We cannot study power by looking only at the resources of those groups which lobby government. We must also examine the resources of their putative rivals and friends. The pluralists are right in their insistence that some groups play a great and significant role in policy formation, but not in asserting that this shows that elitist and statist power ascriptions are invariably empirically refuted. This would only be the case if the pluralists could demonstrate that these policy formations are created despite the resistance of state and elite actors and do not conform by and large with their own policy preferences. Pluralists concentrate their attention upon organized groups and not upon all groups which are affected by any policy outcome (Nordlinger, 1981, ch. 1). However elitists and statists similarly cannot claim that their theories are empirically verified merely because state or elite groups tend to get outcomes they prefer (see Chapters 5 and 6).

Given the bilateral nature of bargaining the calculations of luck and power cannot be quantified because rival interpretations of luck and power will be possible. Consider the following two-by-two matrix:

	Lucky	Not Lucky
Powerful	1	2
Not Powerful	3	4

The elitists claim that there are groups (which together make up a tiny minority of people) who fit in boxes 1 and 2 (usually in box 1) over the vast majority of issues and on the most important issues; and groups (who together constitute the vast majority) who fit into boxes 3 and 4. There is a large degree of overlap in policy fit. Many people in the latter groups might agree with the policy outputs – they are in box 3. Many powerful people are lucky and so do not need to act to get the policy outcomes they desire – they are in box 1. An early pluralist position (by which I mean Dahl, 1961a – at least as he is so often represented by critics) needs to maintain that all (or at least most) people are in all four boxes in different issue-areas. They can agree that some people might end up in box 4 more often than in box 1, and some in box 1 more often than box 4, but must maintain that there is not a systematic bias and this does not occur for the vast majority of people over the most important issues. In other words inequalities are dispersed, though there may be a small set of powerless and luckless people. Later pluralism (by which I mean what pluralists maintain now, and perhaps have always maintained) needs to demonstrate that even if people are not in all four boxes in different issue-areas, institutional structures exist which allow entry into box 2 for many people in different issue-areas.

If most people are in box 3 most of the time then they have little power, but they tend to get what they want. Such a situation is perfectly compatible with pluralist theory because pluralists maintain that organized groups will form only when their interests are threatened. Thus if one is lucky one does not need to be powerful, but if one's luck turns then one may need to form organizations and become powerful. Elitist and statist accounts rely upon the assumption that individuals' interests are not being secured even when they do not try to form organizations. I demonstrated in Chapter 3 that this argument is coherent, but being coherent is not the same as being empirically true. Elitist and statist accounts must empirically demonstrate this. How far they have done so, is considered in Chapters 5 and 6. Chapter 7 will then re-examine the ways in which preferences are generated by society.

5

Collective Action and Dimensions of Power

5.1 Introduction

Two sorts of power are identified in this study: (a) 'outcome power', which refers to the ability of a person to bring about or help to bring about some outcome, and (b) 'social power', which refers to the ability of a person deliberately to change the incentive structure of others in order to bring about or help bring about some outcome. Both sorts may involve people working together in coalitions to achieve these ends, for collectives often have powers which individuals lack. We can study the power of the individuals within a collective as well as the power of the collective as a whole. At times it is worth while to think of the collective as a single acting body; at other times we require more fine-grained analysis, breaking the actions of the collective down into the strategies and actions of its individual members. We have also identified 'luck', which has often been confused with power when individuals or groups get the outcomes they want without exertion.

When writing about power of either sort we must always bear in mind the amount and scope of the power claims. This is best done by studying the resources and aims of the power-holders, together with the resources and aims of those (potentially) in conflict with them and those (potentially) in league with them.

I have suggested that the failure to take full account of the distinctions made in previous chapters has led to many fallacious arguments in both community power studies and national power studies. In this chapter I will provide further evidence of the fallacious arguments and try to suggest how we may overcome them. I will begin by looking at some of the classic studies, for there the fallacies are most apparent, but I will also show in this and the following chapter that they remain in more subtle forms in more recent work.

5.2 Dimensions of power

Lukes (1974) often forms the framework for discussions of power. His three dimensions of power thesis identifies different ways of approaching the problem of power in society. As we saw in Chapter 1, it is essentially a methodological description and critique. Lukes's work is controversial in the real sense of the term. Some well-known and respected academics have described his book to me as 'still the most important and thoughtful book on power', whilst others, equally well known and respected, have claimed it is 'the most appalling work which has held back political science over the last fifteen years'. My own view is that Lukes's work is indeed important. It asks a number of difficult and significant questions which previously had been skated over. Yet I also believe that Lukes's answers to those questions have taken us down a blind alley and concentrated criticism on the wrong factors of those theories he seeks to replace. In fact I believe that Lukes's views suffer from the *same* fault as those he seeks to criticize. He, like them, does not take sufficient account of the problems of collective action, which leads him to commit the 'blame' fallacy. In trying to locate the fault in a different place – the methods of behaviouralism – he created a debate which still rumbles on.[1] Some academics deny behaviouralism, yet their (excellent) empirical work is – as I have defined the term, following the APSA 1944 report – undoubtedly behaviouralist. Behaviouralism as I have identified it is compatible with rational choice methods, with the recognition that individuals may misidentify their own best interests, with institutionalism, and with a form of structuralism which places great emphasis on the relations between people to try to explain why they think and act as they do. Lukes's three dimensions of power take their formulation from their relationship to behaviouralism as he identifies it:

One-dimensional view
Behaviouralist.
Focus on   (a) behaviour,
             (b) decision-making,
             (c) (key) issues,
             (d) observable (overt) conflict,
             (e) (subjective) interests, seen as policy preferences revealed by political participation.

Two-dimensional view
Qualified critique of behaviouralism.
Focus on   (a) decision-making and non-decision-making,
               (b) issues and potential issues,
               (c) observable (overt or covert) conflict,
               (d) (subjective) interests, seen as policy preferences or grievances.

Three-dimensional view
Critique of behaviouralism.
Focus on   (a) decision-making and control over the political agenda (not necessarily through decisions),
               (b) issues and potential issues,
               (c) observable (overt or covert) and latent conflict,
               (d) subjective and objective interests.

Because all three dimensions see conflict as the basis of power Lukes is identifying political power with that aspect which I have called social power and ignoring that aspect which I have called outcome power. The distinctions between the three views are (i) the manner in which individual interests are characterized; (ii) the type of issues to be considered within the scope of power claims; and (iii) the manner in which conflict is characterized.

(i) Lukes's main criticism of behaviouralism is contained in (e) on the one-dimensional view. Here the behaviouralists are accused of equating the interests of individuals with their policy preferences as revealed through political participation. We saw this in Dahl's assumption of two sorts of citizen, *homo politicus* and *homo civicus*, where the latter are reasonably happy with the political decisions made on the issues the New Haven researchers studied. However, our theory of action demonstrates that we cannot begin to understand human behaviour unless we understand human beliefs and desires, or understand desires unless we understand beliefs and actions. In other words, Dahl assumed something about both the beliefs and desires of *homo civicus* from their actions alone. Individuals may abstain from politics not because they are happy with the results their political masters achieve but because they feel and (to some extent therefore) are powerless. The costs of taking part are not worth the expected benefits. Held (1987, 182) has identified five different ways in which such cost-benefit calculations may lead to ready compliance. First, the public is coerced; that is, they feel that they have no choice in the matter. Second, they comply since it is traditional to do so; that is, they have never thought about not complying. Third, they are apathetic; that is, they cannot be bothered to try to change decisions. Fourth, they may acquiesce pragmatically; that is, although they do not like a situation, they cannot imagine its changing. Finally, moving more towards agreement, they may accept conditionally, for whilst they may be dissatisfied they feel that in the long run acquiescence may be to their own benefit. Two other sorts of compliance – normative agreement, where individuals comply because they think they ought to, and ideal normative agreement, where individuals cannot imagine the situation's being better than it is – complete the full range of acquiescent acceptance of political decisions.

We can recognize the strength of these criticisms of the New Haven and other pluralist studies without making any assumptions about real or objective interests or about so-called non-decisions or decisionless decisions. Collective action is difficult and its problems may not be overcome (especially) for large groups of people (Olson, 1971). Beyond this we may *also* see objective interests where people make calculational errors or where there are possible states of the world in which someone would be better off but that person does not perceive that possible world from the world in which they are located. But this objectification of interests is not required for Lukes's criticism of the behaviouralism he identifies and it merely muddies the waters within its critique. However, as we saw in Chapter 3, the importance of objective interests should not be downgraded, and I will return to this issue with regard to preference formation in Chapter 7.

(ii) Lukes's second criticism follows from the first. The issues considered to be important are those over which there is controversy within the community as identified by the researchers. There are four criteria described by Polsby (1980, 95–6) for deciding which issues are important. (a) How many people are affected? (b) What types of goods and services are distributed? (c) What is the value of these goods and services? (d) How is resource distribution affected by this process? But as we saw in Chapters 3 and 4, issues which score highly on all four criteria may not be in the news simply because those who want them raised see no hope of success – or, as Lukes (1974, 21) writes, concentrating upon actions in the political sphere ignores

> the bias of the system [and the way it will] be mobilized, recreated and reinforced in ways that are neither consciously chosen nor the intended result of particular individuals' choices.

All issues which affect the interests of individuals need to be brought into the calculation of the structure of power within a community; though not quite in Lukes's way, for the fact that certain issues are not raised may say more about luck than about social or even outcome power as such.

(iii) The third issue is the way in which conflict is characterized and once more this follows from (i) and (ii). Lukes makes a distinction between overt and covert conflict which is 'observable' and latent conflict which, presumably, is not observable. This three-fold distinction is confusing. Potential conflict usually means that people may disagree about some possible state of the world but have not actually disagreed about it. Actual conflict is where people have argued or fought over some outcome; overtly in the public domain, or covertly behind closed doors. However, conflict in the sense of incompatible preference orderings is actual (in that it actually exists) all the time even when people have not argued or fought over an issue or even realized that there is an issue over which they disagree. A group of people who share an interest in some outcome, say R, will still be in conflict if they also have interests in some other outcome, T. Remember, in the Prisoners' Dilemma each player has the preference ordering $\{T > R > P > S\}$. As individual interest was defined in Chapter 3 (Section 3.6) it is in both players' interests to reach outcome (R, R), yet one player prefers outcome (T, S) and the other prefers outcome (S, T). However, given the rationality assumptions and the structure of the game, neither of those two outcomes is feasible. Thus the players have a common interest in producing outcome (R, R), but despite that common interest there is still inherent (and actual) conflict. Wherever groups share common interests and there are freerider problems this conflict exists. This is actual, not potential conflict, though it may potentially lead to a struggle. (It may be overt or covert but that is neither here nor there.) It is conflict in the sense that the highest-ranked preferences of the concerned parties are incompatible. Such conflict is a necessary but not sufficient condition for conflict within its narrower meaning of actual struggle (Blalock, 1989, 7) but this broader meaning is required for the necessarily counteractual analysis of power. This conflict within overall cooperation is captured in Harsanyi's and later bargaining models where coalitions attempt to bring about outcomes which are in their members' general interests, though not all members may be in agreement over the particular form taken.

This fact of continual conflict because of the incompatibility of preference orderings within groups does not rely upon the simple self-interested assumptions governing individual behaviour with which I have been mainly concerned. Groups form to bring about outcomes in part because of their preference orderings and thus these differences are important even where people do not argue or fight or are not even cognizant of the conflict within the coalition. Collective action problems are not simply overcome by assuming altruism (Frohlich, 1974; Margolis, 1982). Altruists may still order preferences differently from each other, and there is actual conflict whenever preference orderings differ. Some communitarians like to imagine societies where there are no individual variations within preference orderings, but this assumption is far too heroic for my taste and *any* difference in preference ordering is conflictual.

5.3 The 'political power' or 'blame' fallacy

These problems with the decisional models of political power led Lukes and other theorists into new ways of looking at the power structure. However, because they did not take into account the implications which flow from the collective action or freerider problem, even where they recognized it, they were led into committing the 'political power' or 'blame' fallacy. The blame fallacy asserts that, because something did not turn out as intended, there must be someone to blame. For example, when a plane crash occurs there is a tendency to ask, 'Whose fault was it?' Was it pilot error, or a mistake by the ground-staff or the designers or the engineers? Now, whilst there may well have been someone who was negligent, it is not necessarily true that there *was* someone who was negligent. Even if the crash was due to a design fault, it does not follow that there is someone who may be blamed for this. The fault may be due to something which has never previously been realized and, given technical knowledge at that point in time, no one could be expected to have realized it until the part failed and that failure was investigated. The political power fallacy has a similar structure. It suggests that the fact that actor A is powerless to bring about some outcome x implies that there is another actor B who is powerful enough to stop her. The inference is false. Even if there is an actor B who is powerful enough to stop A from bringing about x, the fact that A cannot do so is not sufficient to demonstrate that B is to blame. This fallacy pervades much of the community power debate, because virtually all the protagonists take insufficient account of the collective action problem. The lack of individual or group power may be analysed by examining the incentive structures of actors without the need to impute Lukes's extra dimensions of power. Powerless groups can be powerless all on their own without there being other powerful groups actually stopping them. I think that the fallacy has been generated because analysts confuse outcome power with social power. They fail to distinguish between the ability of one actor to bring about or help bring about an outcome and the ability of someone else to stop them.

The political power fallacy is committed by pluralists within their very assumptions about interests which we saw in Chapter 2 and Dahl's creation of *homo civicus*. It can also be seen when Polsby (1979, 540) only considers two possible explanations for an issue not being raised within a community: 'Either it is being suppressed or there is genuine consensus.' Far more probable is that neither is the issue being suppressed nor is there 'genuine' consensus, but rather that those opposed to the status quo either cannot afford to raise the issue or can see no point in trying to do so. Polsby seems to think that only the wielding of power by one actor over another (the suppression of the issue) can explain its not being raised: a classic example of the power fallacy in action.

Several community power studies which have been used (not always by those carrying them out) to illustrate the second and third dimensions of power make their points fully without these extra dimensions entering into the explanation. Of the second dimension of power (or what they call the second face of power) Bachrach and Baratz (1970, 7) write:

> Power is also exercised when A devotes his energies to creating or reinforcing social and political values and institutional practices that limit the scope of the political process to public consideration of only those issues which are comparatively innocuous to A. To the extent that A succeeds in doing this, B is prevented, for all practical purposes, from bringing to the fore any issues that might in their resolution be seriously detrimental to A's set of preferences.

They illustrate this by their description of the changed political status of blacks in the Baltimore region between 1964 and 1968 and the strategies used by the ruling elite to modify blacks' demands. They argue that, whilst collective action changed the status of blacks, the direction it took was determined by pre-existing mobilizations of bias. The mayor's appointees ('white notables') to key municipal boards and commissions 'could be counted upon to defend the status quo' (Bachrach and Baratz, 1970, 70) and were justified on the grounds that they were the most qualified. Furthermore, the mayor acted swiftly when the Congress of Racial Equality (CORE) designated Baltimore as a target city during 1966, setting up commissions of a biracial liberal nature designed to pre-empt and reduce the influence of the more radical CORE. This changed the key question from 'how can the distribution of power be radically altered?' to 'what can be done for the poor?' This was accomplished through, for example, the Greater Baltimore Committee (GBC), an association of key figures in industry, which had an impressive list of projects, from rebuilding the central business district to helping in the fields of education, housing and research. However, GBC was not interested in redistributing power and wealth, but rather in improving the absolute well-being of all the city's population. Particularly it was interested in the regeneration of the local economy through its rebuilding plans. Bachrach and Baratz describe these strategies as non-decisions, but to me they look like pretty ordinary sorts of political decision (see also Polsby, 1979, 1980). The point is that the relative lack of outcome power of CORE and unorganized blacks in Baltimore may be calculated by examining their available resources in comparison to those of the 'notables' and the mayor himself. We do not yet need to impute Schattschneider's 'mobilization of bias'. Schattschneider, remember, wrote:

> All forms of political organization have a bias in favor of exploitation of some kinds of conflict and the suppression of others because *organization is the mobilization of bias*. Some issues are organized into politics whilst others are organized out. (1960, 71, original emphasis)

But some issues are not organized out as such; they are *just not organized in*. Prior to black mobilization in Baltimore, the relevant issues were not on the agenda. Once they were, through black agitation, then the threatened community responded to deflect the issues. The success of the original black mobilization as the blacks overcame their collective action problem brought a response from other groups who saw their interests threatened. This interpretation of the events is simpler and perfectly consistent with Bachrach and Baratz's description of Baltimore stripped of their confusing 'non-decisional' and 'second face' language.

A similar example of the blame fallacy is provided in Crenson's (1971) famous study of the pollution issue. This is a good example of the fallacy, since his study was so clearly designed and carefully carried out. Lukes (1974, 42) describes it as 'a real theoretical advance in the empirical study of power' which 'can be seen as lying on the borderline of the two-dimensional and three-dimensional views'. Crenson asks why the air pollution issue got onto the political agenda more quickly in some dirty cities than others. He answers that in some cities the issue was quashed prior to agenda-setting. Lukes believes the study demonstrates that U.S. Steel kept the pollution issue off the agenda in Gary, and thwarted attempts to raise it whilst influencing the pollution ordinances without acting or entering the political arena. Crenson (1971, 69–70) makes similar claims:

> U.S. Steel … influenced the content of the pollution ordinance without taking any action on it, and thus defied the pluralists' dictum that political power belongs to political actors.

The evidence for this non-activity is non-existent, but not because Crenson provides no evidence. He provides plenty – evidence of activity, not non-activity. For example:

> Gary's anti-pollution activists were long unable to get U.S. Steel to take a clear stand. One of them, looking back on the bleak days of the dirty air debate, cited the evasiveness of the town's largest industrial corporation as a decisive factor in frustrating early efforts to enact a pollution control ordinance. The company executives, he said, would just nod sympathetically and 'agree that air pollution was terrible, and pat you on the head. But they never *did* anything one way or the other. If only there had been a fight, then something might have been accomplished!' What U.S. Steel did not do was probably more important to the career of Gary's air pollution issue than what it did do. (Crenson, 1971, 76–7)

On one reading this suggests that the non-decision really was a non-decision. U.S. Steel did not do anything simply because it never got around to it. But this is politically naive. Attack is not always the best form of defence. Public relations executives recognize that often the best way to deal with public criticism is to agree with it, emphasize how difficult it is to overcome the problem, and then promise to work hard at it. Then the strategy turns to blaming someone else. This is exactly what Crenson describes in his book. Nor would it be unduly cynical to suspect that officials of U.S. Steel were working hard behind the scenes to keep the issue out of the public domain. Which is what Bachrach and Baratz (1970, 43–6, and see Chapter 2) originally meant by a 'non-decision'.

In fact the most interesting aspects of Crenson's study show how both East Chicago and Gary overcame their respective collective action problems (CAPs) over the air pollution issue. In both cases a political mover, or entrepreneur, working in the public attorney's office was important (McLean, 1987, 35–6). Crenson says that in East Chicago no one could understand why this individual worked in a public office when he could have had more lucrative employment in the private sector. In Gary the first person to attempt to bring clean air was seen off by the party machine. The mayor did not see any political capital in clean air precisely because the consumers of that air were not mobilized. The second entrepreneur was more persistent; at one point he is described as a man in search of an issue, but he also had a private interest in clean air – he suffered from a respiratory disease. What caused the cities to overcome their CAPs at a different pace were certain contingent political features. In the wider comparative analyses later in the book similar variations between different cities may well be due to similar contingent factors. Crenson never asked the crucial collective action question. Why had the pollution sufferers not organized earlier? To answer this question, we do not need to impute power to U.S. Steel; to do so is to commit the blame fallacy. Note that this does not mean that U.S. Steel was not politically powerful, just that we do not need to impute to it some unobservable power in order to explain the lack of power of the suffering public.

Crenson was not unaware of the collective action problem and late in the book he refers to Olson, but here he confuses two separate issues. In Chapter 5, Crenson (1971) explains in a Downsian manner that political parties are not interested in collective goods because they do not allow sustained brokerage (Downs, 1957). However, this does not itself explain why a suffering public does not organize to try to force the political parties to take note of their suffering. Crenson confuses this with an acknowledgement of collective action problems. For example, when he (1971, 89–90) correctly states:

> The dirty air issue raises the prospects of socially concentrated costs and socially diffuse benefits, hence the concentrated opposition and diffuse support

he mixes together costs and benefits of two different parties. The costs are the costs to industry, the benefits are the benefits to the consumers; but industry's costs do not enter into the public's calculus of action except through job losses, which was Polsby's point against Crenson. The costs to the public are the costs of organizing and mobilizing. Crenson (1971, 90) goes on: 'The poverty issue, we might reasonably argue, involves costs and benefits of just the opposite kind', but the poverty issue faces similar collective action problems to that of the pollution lobby (Whiteley and Winyard, 1984). The point Crenson is making here is that, whilst both poverty and pollution involve a redistribution of resources, the first is a redistribution from many people to a few, whilst the latter is a redistribution from a few to everyone – which is why poverty is more controversial than air pollution. In terms of a simple collective action model, we might expect the poverty lobby to be better organized than the pollution lobby, but less successful once the latter has managed to organize.

Where the pollution lobby is less successful, even when organized, we may begin to look at the opposition acting to stop it.

The difficulty with Crenson's account as it stands is that it still leaves unexplained how it is that large groups of suffering individuals in a democracy do not have the power to overturn industrial interests buying help from well-organized party machines. For example, whilst Crenson (1971, 155) is correct when he suggests that parties maintain themselves by buying specific areas of influence, we need to ask, 'For what do they maintain themselves?' Obviously, to stay in office – but a mass public can put a stop to office-holding, an important pluralist point. Why does the public not threaten the parties over the issue of air pollution? Crenson's book does answer this question but the way in which his argument is organized leaves it open to the Polsby charge we saw earlier, that it assumes groups have interests which they do not try to defend by mobilization. Placing the problems of collective action centre stage in the analysis allows us to make these assumptions and also to reject the much-criticized 'non-action', 'non-event' and 'non-decision' nonsense which has pervaded so much of the debate.

Steel producers are most effective in persuading the public that pollution control is not necessarily in their interests when they exploit the competitive nature of local steel production. If we assume that citizens would sooner have no pollution and a local steel employer, but may well trade off pollution against losing that employer, local communities face a collective action problem, though one whose basic structure might suggest it should be easy to overcome. Assume that local communities order their preferences thus: {steel and no pollution > steel and pollution > no steel and no pollution > no steel and pollution}. The local communities face an assurance game to the extent that they believe local pollution controls have a deleterious effect on local steel production if other local communities do not bring in similar controls. The following captures this fear:

		Pollution Y	Control N
Pollution	Y	1,1	4,2
Control	N	2,4	3,3

Where {1 > 2 > 3 > 4} for each of the players, Y stands for Yes for pollution control and N for No to pollution control. As Y, Y is preferred by both communities the game should result in an optimal outcome. However, the problem for the communities is discovering that this is optimal for both. Lack of information on the preferences of other communities, the lack of trust – some communities may suspect others of trying to gain a competitive advantage by implementing controls at a later date – make the seemingly simple assurance game harder to resolve. If one community brings in pollution control when others do not, it ends up with its least preferred option. For whilst they lose their production they still gain the effects of pollution from other communities. Further there is another wider collective action problem which may work against centralized solutions to the informational problem. Steel producers are not just in competition with each other at the local level but also at the national level where free trade operates. The same sort of assurance problem is faced by nations.

5.4 Gaventa's test of three-dimensional power

Crenson's book was not an attempt to prove the second or third dimension of power but was written with Bachrach and Baratz's work in mind. Another, and in my view one of the best community power studies, Gaventa (1980) was written explicitly to defend and prove Lukes's three-dimensional account. He tries to explain how

> in situations of inequality, the political response of the deprived group or class may be seen as a function of power relationships, such that power serves for the development and maintenance of the quiescence of the non-elite. The emergence of rebellion, as a corollary, may be understood as the process by which the relationships of power are altered. (Gaventa, 1980, 4)

He goes on (1980, 256) to argue that

> Whilst each dimension of power has its mechanisms and uses, it is only through the interrelationship of the dimensions and the re-enforcing effect of each dimension on the other that the total impact of power upon the actions and conceptions of the powerless may be fully understood.

His book is largely concerned with the shaping of the wants and beliefs of the people of the Clear Fork Valley in Appalachia by their employers – a British-based mining corporation. In the story as Gaventa sees it, not only does the employer deliberately shape the beliefs and desires of the people, by setting up incentive structures, controlling the press and local politics, but power itself 'shapes participation patterns of the relatively powerless' (1980, 13). Here power itself seems to be an intentional actor:

> As it happens empirically, the first two chapters, dealing with local government and union participation, involve the study of power as *it* serves to maintain quiescence. (1980, 136, emphasis added)

It is my contention, which I hope to demonstrate by a brief review of Gaventa's argument, that the picture he develops of power and powerlessness can be analysed in the much simpler terms of the bargaining model, taking into account the collective action problem.

Gaventa's argument is complex. At each stage he tries to show the operation of each of Lukes's dimensions of power: overt activity, covert activity of the governing elites, and the psychology or consciousness of the mass. The study is also wide-ranging: it examines the history of the area, attempts to unionize the miners, the corruption of the union, and how recent challenges to elite power in the 1970s were beaten off. Underlying Gaventa's entire thesis, however, is the blame fallacy. There is no need to impute a power relation between the elite and the mass in order to explain the latter's lack of power. The powerless can be powerless *per se*. I will try to illustrate this by examining various elements of Gaventa's account.

The early part of the book describes how the company took control of most of the land during an economic boom in the late nineteenth century (1980, 48–58). The overwhelming economic power of the company ensured a political class which supported it (1980, 58–61). Gaventa is not satisfied by uncovering this form of economic power; he also detects the beginning of an ideology created to cement the power structure. Elements of this ideology include a sense of common purpose, the ethic of hard work to attain success and the new capitalist economic order. It was established by 'distorting' information: 'The industrial order was introduced to the mountaineers' society by conspicuous consumption, with an exaggerated demonstration of its benefits' (1980, 63) whilst traditional ways of life were portrayed as deficient. He claims that the local culture was obliterated by replacing old place names with new ones, thus imposing a new identity over the old one. Gaventa writes (1980, 67) that these processes 'were probably not conscious exercises of control', but he argues that they did help to give power to the elite. It seems to me that changing place names is straightforward manipulation. Gaventa also describes other forms of direct exercises of power. Challenges to elite interests were defended by the judiciary and local politicians with direct coal interests. He describes 1930s battles between the mine-owners and the United Mineworkers of America and the National Miners' Union which were overt power struggles. Two journalists sympathetic to the workers were shot. The rise of unionism and then open conflict clearly demonstrates a clash of interests. Why did the miners' interests not come into the open before? Why did they apparently disappear afterwards and mass docility take over? For Gaventa, the third dimension of power is required to answer these questions. However, his account also suggests an answer which is completely in keeping with a rational choice-based account.

A good test of differential attitudes comes from comparison between the reactions of two different communities to the industrial conflict of the 1930s: the quiescence of the Yellow Creek miners and the militancy of those nearby in the Jellico area (1980, 72–80). Gaventa suggests that the contrast rests upon the different consciousness of the people in the two areas – individualism in Yellow Creek, class consciousness in Jellico. However, it is just as reasonable to suggest that both groups of miners faced similar collective action problems but that the Jellico miners overcame theirs. Explaining why may be done at two levels.

First, there may have been a contingent factor for which little or no general explanation need be offered. In the same way as the people of East Chicago may have been luckier than the citizens of Gary, the Jellico miners may have been lucky that their leaders were more dynamic than those in Yellow Creek. Their success might just be contingent upon the arising of political movers in Jellico rather than in Yellow Creek, and some evidence for this exists. Gaventa (1980, 93) mentions one miner in Yellow Creek who said of this time, 'Well, I guess we could have gotten power then, there were so many of us, but we just didn't think of it.' The only general analysis which may be offered of such individually contingent explanation is that statistically we would expect different publics to overcome their respective collective action problems at different though not too dissimilar rates (Crenson, 1971, provides evidence of this).

Second, we may give an explanation couched in terms of the differences between the two communities. Gaventa gives some evidence as to why the Jellico miners were more able to overcome their mobilization problems. He tells us (1980, 47) that the Yellow Creek area grew in population very quickly and had many transients, whilst the Jellico miners were more integrated. Furthermore, a reputation for militancy and solidarity is self-fulfilling and self-perpetuating as group members learn to trust each other (Lange, 1984, 106). This too can help to explain why the Jellico and Coal Creek miners stayed out longer than the Mingo ones in the General Strike of 1894 (1980, 79). Gaventa (1980, 255) well describes how past actions create consciousness:

> Continual defeat gives rise not only to the conscious deferral of
> action but also to a sense of defeat, or a sense of powerlessness,
> that may affect the consciousness of potential challenges about
> grievances, strategies, or possibilities for change. Participation
> denied over time may lead to acceptance of the role of non-
> participation, as well as to a failure to develop the political
> resources – skills, organization, consciousness – of political
> action.

Furthermore, and I think far more importantly than Gaventa recognizes, the troubles of the Jellico miners were different from those of the Yellow Creek miners. They were troubles associated with a specific problem arising at a particular time which directly affected the interests of the miners, viz. the importation of convicts to work in the mines. Collective action problems are always easier to overcome in relation to such issues than with long-term threats (Taylor, 1987), though, once organized, the functional groups are often more efficient (Newton, 1976). Further, threats to what one already has seem more real than possibilities of future improvements of the same value (Tversky and Kahneman, 1981). These are rival hypotheses to Gaventa's, the existence of which demonstrates that he has not proved his point. Explanations of differential rates of mobilization are intellectually more satisfying than offering merely contingent reasons. We should look for structural differences, though we ought not to forget the contingency of human affairs – sometimes the answer lies merely in who happens to be in the right place at the right time.

Of course, beyond these factors affecting the outcome power of the Yellow Creek miners, one of the biggest obstacles they faced was the deliberate attempts of the Company to stop them mobilizing. The Company had greater control here than in Jellico, owning more land, and controlling even more labour and housing. Again, however, this social power can be explained by the Company's relatively greater bargaining resources and does not require us to adduce extra dimensions of power. We may assume that decisions rather than non-decisions were made to ensure that certain laws were not enforced (Gaventa, 1980, 81) and we do not need to suppose that ideology was inadvertently created by the powerful; they can deliberately encourage an ideology that suits their purpose (as Gaventa's later evidence suggests). Whilst ideology may also arise from the way in which a community is structured, that is then luck not power, though of course the form the ideology takes affects both the outcome and the social power of everyone (see Chapter 7).

In fact evidence for the collective action problem can be found throughout Gaventa's book. Dismissal from the Company, not only for oneself but for one's extended family, and loss of housing with little or no hope of future employment anywhere in the region are examples of the great bargaining power of the Company, not of second faces of power (1980, 87–90). There was little hope of Exit for the people, high costs of Voice (1980, 87) and therefore a Loyalty of sorts, quiescence, was all that remained. These high costs of action are not counterbalanced by its immediate benefits, which seem low. This has little to do with the responses or expected responses of the powerful (1980, 91). Gaventa seems vaguely aware of this when he writes of the 'power field' (1980, 151, 257) within which people operate, though he does not explain this metaphor. Perhaps it can be seen as the structure of a collective action problem.

The development of ideology in the Middlesboro region is well-documented by Gaventa, his direct comparison between the news reports of the local newspaper, the *Middlesboro Daily News*, and those of disinterested papers outside the region (*New York Times* and *Knoxville News Sentinel*) showing how the press can help to shape community attitudes. Only the naive now think otherwise. At times of potentially escalating challenge the local press directed attention away from the issues of unionization and unemployment to emotional ones of patriotism, moral behaviour and religion. Importantly the communist organizers of rebellion in the 1930s failed in their attempts to replace these ideological symbols of the mountain people with new ones. The communists frightened off many potential supporters with their atheism:

> Whilst for the communists the church was the 'opium of the masses', to the miners of the coal camps ... it had been the only form of collective organization allowed. (1980, 115)

The workers lost the industrial conflict of the 1930s, not because of some collective lack of consciousness, or lack of understanding of their real interests, but because their opponents had greater resources, financially, politically, institutionally. The capital side of the battle was much better resourced and organized than the union side (Gaventa, 1980, 108).

The United Mineworkers of America (UMWA) was not in any sense a democratic organization and became dominated by its leader, John Lewis. In Gaventa's account it too begins to work against the interests of the masses and controls them just as effectively. The workers who fought for the right to join a union were then kept in check by the one they had joined. The controls exerted by the union and its leading officials to hold off any challenge look like quite standard uses of political and corruptly political weapons. Jock Yablonski, a disillusioned member of the Executive of UMWA stood against Tony Boyle for President of the Union in 1969.

Gaventa suggests a multitude of reasons why miners should have been unhappy with their union, from its lack of lobbying for better health and safety in the mines and its mismanagement of pension funds to outright corruption of its officials (1980, 172–8). Yablonski was not successful: facing a strong campaign to discredit him, his supporters were physically beaten and he was eventually murdered.

The murder of Jock Yablonski and the corruption of the union are, incredibly enough, supposed by Gaventa to support the second dimension of power. Conspiracy to murder and the beating of Yablonski supporters fit into the Bachrach and Baratz definition of 'non-decision' comfortably enough only so long as we retain the Eastonian systems metaphysics with its particular neat division of that which is inside the political system from that which is outside.

The assassins were paid by money raised by district organizers and 22 pensioners who were not initially aware of the intended use of the money. The fooling of the pensioners and their willingness to perjure themselves later when the crimes were being tried is used by Gaventa as an example of both the second and third dimensions of power. However again another interpretation begs consideration from the evidence which Gaventa displays. At first, the pensioners were merely fooled. They were aware that their contributions were in some sense illegal (being 'kickbacked' to finance a campaign against William Bell of another 'company-controlled' union who was running for a judgeship), but had not realized they were for the murder of Yablonski (1980, 182). Later they lied to investigators, for they feared a loss of their union welfare and pension rights. It is true that this fear seems to be an example of the law of anticipated reactions rather than a result of direct threats to their benefits but this law is sometimes the result of power and sometimes of luck, but to explain it does not require dimensions of power additional to those contained in the bargaining model.

Support for Gaventa's defence of the third dimension of power may be provided by the results of District 19, which nevertheless voted heavily in favour of Tony Boyle rather than Yablonski's successor, Miller, at the rerun election despite the murder, the arrest of District 19 officials, Boyle's indictment for corruption, and the obviously corrupt first election, which had led the federal authorities to demand a rerun. Gaventa (1980, 194) says:

> votes for Boyle reflected neither consensus nor coercion, but a socially constructed assessment of the costs and benefits in taking sides in what was perceived by the powerless to be inconsequential conflict amongst the powerful.

In other words, the members made a rational decision based upon the expected benefits of victory for the Miller/Yablonski side winning, that is, virtually none, and victory of the Boyle side, that is, virtually none, and the costs of voting for one side or the other. The costs of not voting for Miller were zero, it may be hypothesized, but the costs of not voting for Boyle were, as Gaventa demonstrates earlier, possibly enormous. Boyle got their votes because he was socially powerful through the resources he had available. Ideology was also important, but Gaventa demonstrates how the union deliberately inculcated that ideology.

Indeed, Gaventa provides good evidence of a lack of interest in politics more generally and union politics in particular, for the people he studied saw little difference in the elites who ran for election. Participation is not rational if costs exceed benefits, and if what Anthony Downs calls 'party differential' is low then the benefits of a win for the side one favours are not much greater than the benefits concomitant upon a victory for the other side (Downs, 1957, 39–40).

Gaventa demonstrates how the people of his valley began to challenge the powerful by pursuing strategies to overcome their mobilization problems. A Granada television programme gave them a medium over which they had some control and increased interaction between members of the group, which Truman (1951) argued is important to mobilization, causing individuals to realize that they were not alone. He also movingly describes how unsuccessful they were in fighting the multinational company controlled from London. Exposés rarely translate into effective protest unless other groups can also be mobilized to keep pressing, and that is difficult across national boundaries (Gaventa, 1980, 246). He describes 'limit situations' (1980, 209) within which action is thought to be possible. Individuals' perceptions of what is possible are altered through participation as they see the possibilities of acting collectively to secure their interests. As cooperation proceeds the possibilities inherent in collective action become apparent and a culture of cooperation begins to develop:

> As actions upon perceived limit situations were successful, more participation occurred, leading to further action. In a concrete situation an interrelationship begins to be seen between participation and consciousness, so that one becomes necessary for the development of the other in the process of community change. (Gaventa, 1980, 213)

This collective protest was unsuccessful when the state and federal bureaucracies were seen to work against them and the opposition forces were able to mobilize in response. The lessening of protest was consequent upon some intimidation as the costs of collectively acting were increased.

Gaventa is right in his claim that many community power studies have not got to the root of the difficulties that groups find in mobilizing to press for their own interests. He is right that many factors, including prevalent ideologies and belief systems, work against individuals' perceiving what is in their best interests and realizing the potential for their own collective action. He is right that historical analysis is vital to understanding contemporary political processes. But he is not right that we need a model of two-dimensional and three-dimensional power as Lukes defines these terms in order to demonstrate these factors. His book is a brilliant account of the problems of people in a part of Appalachia. He has shown researchers how to go about community power studies, which require careful historical analysis as well as opening out the area of research to the national or even international arena. But he has failed to recognize that the starting-point for study is the collective action problem, and the openly used legal practices and more hidden corrupt practices used by the powerful to keep that mobilization problematic. In fact Gaventa's study does not have to be non-behavioural, non-decisional or multidimensional.

5.5 Local state autonomy and the growth machine

In Britain one of the best community power studies is Newton (1976). There is some dispute over whether or not this is a pluralist study, with Newton and Polsby giving different interpretations of (largely) the same evidence (Newton, 1976, 1979; Polsby, 1979). Part of this dispute centres around Newton's overly egalitarian interpretation of pluralism. His own argument focuses on the independent power of local state actors, both elected and appointed, and the ignorance of local politicians on some of the issues studied. Newton is also interesting from my point of view for his misinterpretation of Olson's logic of collective action:

> Mancur Olson argues that small groups are likely to be both more effective and more active than large ones. The contrary hypothesis is argued here, namely that large groups will be more active because the larger the number of members, the greater the resources of voluntary labour, the greater the likelihood of public authorities taking action which affects the interest of some members, and the greater the likelihood of the organization having a critical mass of members who will want to react. As it happens, the correlation between membership

> size and political activity is insignificant so neither hypothesis
> is supported. (Newton, 1976, 43)

However, Newton uses the term 'group' differently from Olson's official definition in developing this argument. Group for Olson, as in standard group theory, refers to any set of people who share some common interest; but in Newton's 'test' the word stands for 'organized group'. (In Newton's defence it must be pointed out that Olson confuses the two uses too.) Olson's argument focuses upon the greater coordination difficulties facing large groups; Newton's on the legitimacy of larger groups and the likelihood that some will react. In fact the resources of a group are far more important. Larger size may be helpful here to help meet organizational costs and to fund effective political action, but the wealth of group members is likely to be more important still (Goodin and Dryzek, 1981). Far more important is the quality of the officers and the nature of the claims of the group. Those groups which are not organized are the most powerless, and the fact that organized groups have differential effects upon outcomes is quite compatible with pluralist arguments, as Polsby makes clear.

Newton describes how some organizations are considered to be legitimate by local state officials whilst others are outsiders. The nature of the claims of groups, how closely they fit the aims of state officials, is an important aspect of power relations which I will consider in greater depth in Chapter 6. However, in terms of public profile, 'legitimate' groups are often more hidden and their claims tend to be processed by appointed officials within the bureaucracy; less orthodox and more controversial organizations would have their claims passed on to the relevant council committees of elected politicians (Newton, 1976, 67). It is this which leads Newton to suggest an autonomy of the local state (cf. Jones, 1969, 280–81; Bealey et al., 1965, 339–40, cited by Newton, 1976; see also Sharpe and Newton, 1984). This autonomy occurs largely as a result of local bureaucrats defining for themselves which groups are orthodox or legitimate and taking note largely of those. Thus the success of organizations depends as much upon their interests being convergent, or at least not divergent from those of the local state officials. The less orthodox groups have a much lower probability of succeeding. Often the power of the 'legitimate' groups depends upon their close links with the bureaucracy and their role as information providers and processors. Newton also argues that rarely were lobbying organizations aware of organizations lobbying contrariwise.

The ignorance of local elected officials is demonstrated by Newton's account of both the housing issue and race relations – as well as a considerable amount of racism, which one suspects would be more carefully hidden these days. A considerable number of councillors, even those sitting on the relevant committees, were unable accurately to state the length of the city's waiting list for housing, nor had they any idea of the size of the black population. The most interesting aspect of these intertwined issues was how the council unwittingly encouraged the creation of immigrant ghettos by its housing policy (Newton, 1976, 194–221). It created a five-year residency requirement for council houses, when 40 per cent of the housing stock was in council hands and 40 per cent privately owned, forcing new arrivals into the 20 per cent of housing available for private rental. It also used the 1962 Town and Country Planning Act and the Birmingham Corporation Act of 1965 to contain the spread of multi-occupancy housing, which was also in the main areas of black residency. Certainly, here councillors wielded power over the black population, aiding and abetting ghettoization, though how deliberate this was is moot (Newton, 1976, 219). Either way, a deliberate ploy or more likely the unintended consequences of action, the structuring of the choice situation of the black community was a result of the actions of councillors and not the action of structures as some would have it.

Many of the local power studies in both the United States and Britain have seen power in the hands of either local elites or, more recently, in less localized business elites. Friedland (1982) argues that the interests of business within urban communities are looked after because of local dependence upon its investment. Locational flexibility of private investment increases with larger and more geographically extended businesses and leads local politicians to plan to attract and keep investment locally. Businessmen do not need to intervene through any organization in order to promote their general interests. They may be seen as lucky rather than powerful, for their interests are also seen by the local community as being in the general interests of the community. Friedland discovered that cities with powerful corporations were very responsive to any localized conditions likely to restrain local economic growth. Thus postwar urban renewal occurred in communities with growing companies and where retail economies were strong, rather than in local economies which were weak. The business which controlled investment also controlled the conditions which enabled the success of urban renewal. We do not need to assume 'voiceless power' (Friedland, 1982, 2) in order to explain this. In making this assumption Friedland is forced to try to locate the power of individual businessmen within the organizational structure of which they are a part. Here the source of a businessman's power derives from his position in an organization. This latter statement may well be true, but it is not required in order to explain the effects Friedland discovers. We may avoid his search for organizational participation in local decisions. When business is lucky it has no need to exercise the power it may have. It does not need to intervene. This is not voiceless power, merely luck.

Friedland's analysis requires an overly strong assumption of capital locational fluidity and we cannot argue from his analysis that local business is politically powerful. Rather we can note the correlation with local economic success and urban renewal and suggest that the former facilitates the latter. That urban renewal is in the interests of local business has never been in doubt, but then it is in the interests of most of the local community too. Rather we might sooner conclude that business and the local community (in the main) are lucky rather than powerful when those facilitating conditions obtain.

In Britain, Saunders's (1979) study of urban development in Croydon reaches similar conclusions. Here Saunders argues that the biggest gainers were larger businesses coming into Croydon rather than the smaller businesses already there. The elected Town Council itself was dominated by the former and Saunders finds little evidence of substantial outside lobbying to bring about particular development. He relies upon assumptions of personal contacts at the network of local clubs, chambers of commerce, and so on, and the assumption of a common ideology of urban renewal amongst all businessmen in order to explain the development (see also Soloway, 1987, for an example of 'old boy networks' in Dahl's New Haven). Those who lost out in Croydon's building plans were working-class residents who would have benefited from the high-density housing which was successfully resisted by the organized middle class. However, the important point which Friedland in America and Saunders in Britain have in common is the significance of constraints upon local actions by fear of the effect on local investment decisions made largely by non-local businesses. In Chapter 6 we see similar arguments apply at the national level between governments and multinational companies.

A particular policy study – the development of high-rise housing in Britain – led Dunleavy (1981a) to stress similar factors. The most important outside influence was the cost controls exerted by the Ministry of Housing and Local Government. Dunleavy provided evidence of strong pressures upon the Ministry from the construction industry, but also of a culture of high-rise buildings fostered by professional architects jumping onto the bandwagon of novelty and with ideas of containing the expansion of the cities. However, the idea that this study supports a structural, rather than a behavioural account is overblown and incorrect. The behaviour of the actors can be understood in terms of their beliefs and desires, in part structured by the influences mentioned. Again, we do not need to argue that those who gained through high-rise building had power, nor that those who lost did not. The power of the construction interests can be seen in the resources with which they lobbied. State actors within the Ministry had little to gain or lose with high-rise rather than low-rise dwellings; rather they were indifferent to the decision and so were swayed by the professional advice they sought (Dunleavy, 1981b).

The importance of the fixed capital 'land' in community politics has been emphasized in the 'Growth Machine model' in the United States (Molotch, 1976, 1979; Logan and Molotch, 1984, 1987; Domhoff, 1983, 1986). The one aspect of local communities that ties all studies together is the physical aspect of their geographical extent. Land is a fixed capital in which all property-owners have a shared interest. All property-owners want the value of land in the community to rise. New development and local economic growth are the surest way of achieving that rising valuation. The Growth Machine model posits that local power is structured around land-based interests. This secures a common interest between ordinary home-owning householders and capitalists or *rentiers* owning larger tracts. All these people have a common interest in providing the right sort of conditions to attract outside investment, though they may have conflicting interests over exactly *where* within the community they want development. Organized capitalists initiate development within the community but local government will back such development, seeing it in the interest of the community to which it will be sold, not only to home-owners but also to the working community for which it will secure employment. Logan and Molotch (1987, 21) argue however that benefits are *skewed* from the general public to the *rentier* groups and their associates. Furthermore they suggest that exchange-values are favoured over use-values and the intensification does not necessarily achieve growth. I will return to this point below.

The Growth Machine literature sees this power being structured around elite figures who will 'intertwine' among boards of directors of local companies.

> [T]he central meeting points of the growth machine are most often banks, where executives from the utility companies and the department stores join with the largest landlords and developers of the boards of directors. (Domhoff, 1986, 59)

The Growth Machine literature is associated with elite theory because of its emphasis upon the relationships between these executives of local and corporate companies arguing that these people are the prime cause of local development. Domhoff, for example, argues in his re-examination of Dahl's New Haven study that Mayor Lee was not the prime mover of redevelopment as Dahl maintains. Rather, the local business community met Lee within two weeks of his election in 1953 to 'urge their program on him' (Domhoff, 1986). Domhoff argues that Yale had been negotiating with the city for redevelopment of land for two years prior to Lee's election (1978, 1986). Further, Growth Machine theorists use the evidence of 'redevelopment' being cited in virtually all studies as one of the most important issues facing the community (Hunter, 1953; Dahl, 1961a; Agger, Goldrich and Swanson, 1964; Jennings, 1964; Bachrach and Baratz, 1970; Saunders, 1979; Stone, 1976; Peterson, 1981) to support their model. The local elites would view development as the most important issue, for it is the one which affects their interests most directly. However this emphasis on elite theory as the prime cause of development and hence elite power is not inconsistent with more structural versions of power, as Stone's work (1976, 1980, 1986) demonstrates.

The major theoretical difficulty with the Growth Machine model is that whilst undoubtedly development is, generally speaking, in the interests of landowners it is also, generally speaking, in the perceived interests of everyone in the community (Peterson, 1981). Molotch (1976, 320) describes employment as the 'key ideological prop for the growth machine', but it is also the key welfare prop for the majority of people and can hardly be said to be against their-interests. Whilst the prime movers of development may indeed be landed and capitalist interests this does not really demonstrate the degree of local power that these analysts assume. At the most it provides evidence of their outcome power. They want x, act to get x and get something like x. But (a) whilst businessmen may be the prime movers, this does not make Mayor Lee and his ilk irrelevant to getting development going; nor (b) does it show that they are powerful enough to get their way if the local community is opposed to development. One is not using social power if one acts to produce outcomes which everyone wants. Thus the elite Growth Machine theorists have not demonstrated that these groups are sufficiently powerful to achieve development despite opposition, nor that the business groups themselves have the social power deliberately to change the incentive structures of local communities to favour development. Discussing a number of community studies, the elite theorist Thomas Dye (1986, 46) writes:

> Although machines and reformers differed over which organizational forms should prevail in municipal government and which ethnic groups should get city jobs, they did *not* differ over the goal of economic development. Machine politicians

and their reformer protagonists argued over corruption, patronage, and ethnic influences in city government. But they did *not* argue over economic influence or even social distribution.

The Growth Machine model demonstrates at most the outcome power of landed interests, but at least no more than that landed interest is lucky. What is in the general interests of landowners is local growth, but whilst it may bring them greater material rewards than the propertyless and increase relative inequality between owners and non–owners, growth is supported by (virtually) all members of the community. It is landed interests' luck to be pushing at an open door.

One interesting study which demonstrates both the power of a local council and the independent power of business interests is Bassett and Harloe's (1990) account of the growth coalition in Swindon. Here in the immediate postwar years the growth was council-led. The Labour council, 61 per cent of whose members were railwaymen with close ties to the trade unions, the Independent opposition, the predominant employers – the newly nationalized railway plus new defence-related industries – all backed the 1947 development plan for growth to a community of 80,000–100,000. Helped by national government policies and willingly embraced by private developers, Swindon grew rapidly. Its external environment helped greatly: the fast London rail line and the M4 motorway opened in the late 1960s encouraged new industries into the town to replace the declining railway and other engineering works. Indeed, Bassett and Harloe (1990, 47) suggest that the local council may have played no more than an 'enabling' role by the 1970s. However, the new employment opportunities were not always to the benefit of the local skilled workforce, and many railway engineers left the town to move to Derby, York or Crewe. Whilst growth continued unabated, the central role of the council declined. Indeed, Bassett and Harloe's account of the importance of the council in the postwar period does not really contradict Growth Machine accounts. They argue that the council was able to control growth because it was a major landowner; once it lost this central position, it also began to lose control.

By the 1980s the town began to have doubts about further growth. Public opinion began to suggest that continued development was undesirable, whilst central government controls on local expenditure led to doubts about the local council's abilities to fund further services and local infrastructure. Its land-holdings were declining and the rising price of land and capital expenditure controls meant it was unable to replenish its 'land-banks'. At this time in the mid–1980s private developers pushed for expansion on the northern borders of the town. This land, owned by the developers since the mid-1970s, was prime for development. They intended to provide 9,000 houses for 24,000 people, thus expanding the town by 20 per cent. A public inquiry was held in late 1987, prior to which it appeared that the local councils would oppose development. However, the developers had an important bargaining weapon. First, few people doubted that permission would eventually be granted. Second,

> the developers had offered to contribute towards public service provision. But it was made clear that if the authorities did oppose the plan at the inquiry all such deals would be off and they would lose such leverage as they had to extract such contributions from the consortium. After various discussions and manoeuvres … both councils withdrew their objections to the plan in exchange for contributions to infrastructural and other costs. (Bassett and Harloe, 1990, 55)

All of the Growth Machine studies and most (though not quite all) community power studies demonstrate that development is not just another local issue, but the major issue facing all communities. This should be no surprise: the national economy is, generally speaking, the major issue facing national governments. How far local growth is in everyone's interests is a moot question. If the race for local growth does not really create jobs but merely redistributes them, then perhaps the general public does not really benefit from Growth Machines. Certainly, new jobs in localities do not always benefit the local population, but tend rather to be taken up by immigrants or non-residents. However, the changing occupational structure of a community cannot be simply claimed to be for or against the interests of the local community, any more than it can be simply claimed that technical innovation was or was not in the interests of the machine wreckers during early industrialization (Hobsbawm, 1952).

Local elected officials usually see any employment opportunity as beneficial and tend to back local growth strategies. In fact, competitive pressure among local communities shapes the incentives of locals in favour of development, in the same way that it shapes the incentives of steel-producing communities to be wary of localized environmental control. If one local political community aggressively favours local growth with supply-side incentives for business to locate there, another finds its local employment prospects decreasing. Thus all localities are forced to devise incentives to business to locate in their area. However, if all supply these packages none benefits at the expense of others. Rather each community's efforts cancel out the others' efforts and business locates itself wherever it sees advantages. However, if all local communities pursue growth strategies they will have weakened their local economies through the incentives – either through spending to bring industry, or through local tax concessions. The local communities are in a Prisoners' Dilemma:

		LG X	
		Incentives	No Incentives
LG Y	Incentives	3,3	1,4
	No Incentives	4,1	2,2

where for each local government $\{1 > 2 > 3 > 4\}$ (King, 1990, 279–80, who also points out differences between the British and American cases). The fact that these inter-community Prisoners' Dilemmas exist helps to structure the interests of the local communities towards development in their own areas. Whilst business may encourage this process, as the Growth Machine argues, the interests are structured thus by the fact of community competition, not by the deliberate actions of any actors. The caveat to this analysis is that local growth may not be simply zero-sum. But it is still open to doubt whether local growth strategies add to the positive elements of that sum.

It is not true, of course, that all communities favour development, or that they all favour the same sort of development. Swindon favoured growth in the postwar period, Banbury shunned it (Bassett and Harloe, 1990; Stacy et al., 1975). Teesside tried restructuring its heavy industry in the 1970s; Lancaster tried to improve the small-scale industry and tourism it already had (Hudson, 1990; Urry, 1990). The degree to which local state officials are able to succeed in their aims, the nature of those aims, and the strategies they pursue all depend on local conditions. Careful local study is important. Harding (1990) warns us against a too simplistic account of growth arguments. Whilst property is an important common interest amongst its owners, which may be used as an explanatory tool at the level of generality here, it is a simplification and many conflicts of interest arise amongst property-owners in actual practice. Of course, there are groups who oppose development. Property-owners may oppose some types of development in some areas. Home-owners do not want large factories on their doorsteps even if they do want them in their general locale. A mark of citizen power is how far they can determine where development takes place. This power has been demonstrated where development within certain areas is stopped and this is where pluralist analysis is seen at its strongest. Of course, politicians of all political persuasions tend to favour development in run-down areas, and it is the ill luck of those who do not favour such development to be so hounded (see Chapter 7). The real battle of urban renewal is over what form it should take. Crudely we may say that, if poor urban housing is replaced by high-rise office development, then landowners and developers have benefited; if it is replaced by better urban housing then poorer community associations or the middle class have been successful. Either way elected officials who hold the state resources finally decide, influenced by the relative power of the contending groups.

Developers favour green-field development for its higher profitability, but increasingly face strong environmental opposition. Some studies show that where such resistance to development occurs it is slowed or stopped. Indeed, there is a vast haul of studies of single-interest organized groups fighting development around broadly environmental issues with which pluralists have made their case (for example, Kimber and Richardson (eds), 1974; Lowe and Goyder, 1983; Richardson and Watts, 1985). Studies in America have shown that successful lobbies against development tend to be either student-led or middle-class in origin, but then the greater mobilization power of these groups is of no surprise. One of the best single-issue studies (Kimber and Richardson, 1974) demonstrates how much more effective the middle-class Wing Airport Resistance Association was than its predominantly working-class Foulness rival in fighting the proposed site of the third London airport.

How far do the elite studies support elite theory and undermine pluralism? They support elite theory in so far as it argues that there are groups of people who have greater resources than others and get their way on a number of issues that affect their interests. They support elite theory in that these groups are made up of very small numbers of elites who can get their way with comparatively little opposition since most other groups perceive these elite aims to be in their interests too. They support elite theory to the extent that it relies upon elites getting their way because they are lucky, perhaps systematically lucky. They support elite theory to the extent that it suggests that, for all sorts of reasons, the wealthy benefit proportionally more often. It does not undermine pluralism in so far as pluralism is not in disagreement with these aspects of elite theory. Pluralism requires only that groups have access to channels of power and influence to make their voices effectively heard. Indeed the elite theorists themselves demonstrate that groups do have influence in policy arenas outside the growth issue.

Poorer groups in society do not, in the main, oppose local development, though other issues – particularly allocational and redistributional ones – may be more important to them. However, redistributional policies are more often handled by central rather than local governments in both America and Britain. Allocational issues are generally of more interest to pluralist researchers, for they more directly show the clash of interests of different organized groups in society. Groups undoubtedly do have some power to influence the allocation of local goods and services such as police, fire protection, education, and so on. It is here that the pluralist case is most easily made and defended (Peterson, 1981). Elitists do not generally dispute this but claim that these issues are not the important ones for the real power structure. However, the central claim over the local economic issue merely reflects broader arguments at the national level. Certainly, local politicians, local party machines and local bureaucrats support local growth schemes and even initiate them, as some 'new urban left' councils have done in Britain or Mayor Lee arguably did in New Haven; and, of course, they are in the interests of local capitalists and *rentiers*. However, this does not demonstrate the power of the latter, only their luck.

5.6 Conclusions

In this chapter I have given a brief review of some of the best-known community power studies to demonstrate infelicities and fallacies in their arguments. Essentially the same fallacy operates from both sides of the debate. Groups do not have to face explicit opposition in order to be powerless, for first they have to overcome their own collective action problem. How that problem is structured may depend upon deliberate actions of powerful individuals and organizations in the past but not necessarily upon action on the part of the powerful today. Rather those who benefit from others' collective action problems are lucky. Once mobilized, these groups may then face opposition from those whose interests are threatened – as traditional group theory supposes (Truman, 1951). Often the management side of the capital/labour divide organizes second, in response to the organization of the labour side (Walker, 1983). Similarly, in communities the citizens organize first, and then those whose interests are threatened respond. We cannot blindly state that because someone's interests are furthered they are powerful, they may just be lucky – though of course it is possible to be both powerful and lucky. The test of their power is how they respond to the mobilization of contrary interests. The evidence prior to that response is the resources they are known to have at their disposal. In the next chapter I will broaden the analysis to the national level, where we will see many similar arguments operating. I will consider at greater length the argument that the state – both national and local – has greater autonomy than pluralism traditionally allows. This older-fashioned argument – that it is the state rather than groups which has power – is not well formulated, though it contains a great deal of truth. I will argue that it is groups which are powerful, but that state groups are often very powerful because they have state resources to bring to the societal bargain, though they too must operate within constraints. Constraints are exerted both by organizations representing the interests of citizens and by those that represent more diverse functional interests.

6

State Power Structures

6.1 Introduction

We have seen that many fallacious arguments are utilized in the power literature on urban communities. The fallacies appear on all sides of the debate in pluralist, 'radical' and elite theory. These same fallacies often reappear in the literature on power structures at the national level. I will not give a detailed critique of particular works in that literature but rather will broaden the account to consider different models of the policy process. As the dominant and most criticized literature is pluralist, I shall begin by looking at pluralist accounts.

Pluralism has been broken down into different sorts of categories such as 'neopluralism' and 'reformed pluralism' (Dunleavy and O'Leary, 1987; Smith, 1990c) but this is not the best way of organizing the discussion. Pluralism, like liberalism, marxism, and so on, is a tradition with many proponents and a long history. Not all pluralists agree with each other on all points, and like all traditions pluralism has evolved over time. A charitable approach to criticizing a tradition is to make the best case for it and totally to reject it only if that best case is found wanting. If some elements of pluralism can be retained then we may as well work within the tradition even if it ends up far removed from the original theory. I believe that it is hard to reject interest group pluralism as a theory of society since it is almost trivially true. However the worth of the theory as an organizing principle derives from the meat that other so-called competing theories may put on to its dry analytic bones.

Interest group theory is really very simple. It argues that society may be broken down to a set of analytic categories. These categories constitute interest groups, which are defined as sets of individuals who share some common interest. I take 'interest' to be as defined in Chapter 3. If the common interest is threatened then the group may organize to defend that interest. The theory as originally propagated by Arthur Bentley (1967) was devised, partly, as an analytic theory to counter marxist class theory. Its strength over class theory is that it recognizes the many cross-cutting cleavages through the individuals and groups of individuals which make up society. Its weakness as a theory in comparison with class analysis is that the common interests of group theory do not translate so easily into action as the commonalities within class theory. However, simple class theory has proved to be empirically false and modern complex class theory (for example, Wright, 1985), which leads to a multiplication of classes, looks ever more like group theory, raising similar theoretical problems for mobilization. Further classes within class theory may be subsumed without residue into a specific form of group within interest group theory. Interest group theory is thus an analytic theory which is almost trivially true. The greatest problem for interest group theory is the translation of common interests into common action (Olson, 1971); modern group theory, however, recognizes that the mobilization question is the most important one. How far organizations correspond to all the interest groups which exist, and how far those organizations represent the interests of the groups they purport to represent is the key question for normative pluralism (Dowding, 1987).

Pluralism, as a tradition, grows out of interest group theory. Pluralism suggests that at all levels of government policies result from a complex interplay of pressures deriving from various interest groups, both organized and, through the rule of anticipated reactions, unorganized interest groups. It does not need to argue that all interest groups are equally represented nor that all are equally powerful. Rather it argues that, in pluralist societies, institutions exist which allow for representation and pressure from all 'legitimate' (Dahl, 1986, 182) interest groups. Representation may be seen as some function of (a) intensity of preference, (b) size of membership, and (c) degree of mobilization. The third category, however, depends to some extent upon the first two, upon the relative wealth of group members (which affects relative costs) and upon the disincentives towards mobilization created by members of groups with contrary interests. However, even if we argue that these problems of mobilization are so great that many groups are not represented, we have not destroyed pluralism *per se* so long as institutions exist which allow for that representation. We may be critical of present institutions and argue that institutional arrangements should be altered to improve the representation of some groups, thereby making the state more pluralist, but we are thus operating within the pluralist tradition.

Similarly we may improve pluralist theory by pointing to important institutional differences between societies in terms of the organizing principles and legal status of different groups within the state or across states. Some groups have a corporatist status and some a more traditional pluralist status. Again, however, these arguments operate within the tradition of pluralism. Elite theory too, whilst in some sense an obvious rival to pluralism, may be seen as putting meat on the bones of pluralism. It is no surprise to learn that elite actors wield greater powers than non-elites and, indeed, given the positionality of formal power locations, this can be described as a trivial truth. Elite theory becomes important when it demonstrates that certain social groups form elite groups which knowingly look after their own interests *to the continual detriment* of the interests of other social groups. It is only a complete rival to pluralism, however, if it shows that pluralist institutions can never give representation to most interest groups. If it merely shows how pluralist institutions are perverted by elites then it may be a critical theory operating within the tradition of pluralism. As such a theory it can suggest ways of improving institutions in order to promote pluralism.

However, two radical approaches to power do suggest that pluralism may necessarily be perverted. The first suggests that the state has its own interests and that it has the power to promote those interests at the expense of other interest groups. The second suggests that the structure of power under capitalism always works in the interests of capital.

In this chapter I will show that a version of the 'autonomy of the state' thesis is compatible with the methods of studying power that this book recommends, and that it is compatible with pluralist tradition. It is important, moreover, for its demonstration that some interest groups have institutional powers beyond those of others; it reminds us that the state is not a neutral cipher as some pluralist accounts suggest (Macridis, 1961; Dunleavy and O'Leary, 1987). I will then demonstrate that the second approach, the structural argument, is compatible with the methods for studying power recommended in this book and does not require us to delve into utilizing the concept of 'structural power'. This demonstration requires, as we saw in Chapter 3, careful examination of the pluralist assumption about interests. The generation of interests is as important to the study of power in society as any other aspect. The demonstration of how these three approaches to power may be examined by rational choice methods will help the emergence of empirically demonstrable differences between them. My aim is to transcend the theoretical disputes in order to bring into sharper relief the empirical differences.

First, I will examine pluralist responses to these two critiques. Pluralists have recognized different policy styles within different policy communities. From these distinct policy communities, they are able to recognize the dominance of certain sorts of groups in certain policy areas and to study the way in which different institutional factors lead to different policy styles. However, I will argue that not all policy communities can be seen as pluralistic and we must be careful not to turn the trivial analytic interest group theory into a trivial pluralist argument.

6.2 Policy communities

An important aspect of the pluralist literature has been the development of the idea of 'policy communities' and 'policy networks' (see the reviews in Rhodes, 1990; Jordan, 1990a, 1990b). A policy network has been defined as a

> complex of organizations connected to each other by resource dependencies and distinguished from other ... complexes by breaks in the structure of resource dependencies. (Benson, 1982, 148, cited Rhodes, 1988, 77)

Rhodes (1988, 77–8) argues that these dependencies vary along five key dimensions:

1. constellation of interests – the interests of members vary by some function of territory, expertise and economic or social position;
2. membership – between public and private sector, political or bureaucratic elites, professions, unions or consumers;
3. vertical interdependence – the interdependence of members varies across networks;
4. horizontal interdependence – relationships between networks vary in the degree of their interdependence, networks may conflict or cooperate or be indifferent to others;
5. distribution of resources – members of networks control different types and amounts of resources.

A policy community is a set of networks which are characterized by a stable relationship, the continuity of a legitimized and restricted membership, vertical interdependence based upon shared responsibilities for policy outcomes and largely insulated both from other policy networks and public or Parliamentary scrutiny. This suggests that only certain privileged groups have access to the policy process and are thus influential. Organized groups outside the process have little or no power over the eventual outcomes. This viewpoint may be seen to lie between state autonomy views and the pluralist picture of Dahl, though it is not far removed from Dahl's proviso

(1986, 180) that 'a group has to be seen as legitimate in some sense in order to gain entry into the political system'. Entry into the political community is not automatic for any organized group. But Dahl (1956, 145; see also 1986, 180–81) also says that under pluralism

> there is a high probability that an active and legitimate group in the population can make itself heard effectively at some crucial stage in the process of decision. ... When I say a group is heard 'effectively' I mean more than the simple fact that it makes a noise; I mean that one or more officials are not only ready to listen to the noise, but expect to suffer in some significant way if they do not placate the group.

In other words, even when groups are not able to get their way, they can still inflict costs upon state officials. Metaphorically they bring some weight to one side of the power balance. The policy community picture suggests that some communities are relatively closed, with only certain organizations allowed a hearing, whilst the noisy ones outside do not make public officials suffer in any significant way. Other policy communities may be more open, but in neither Britain nor in the United States is the system of government as permeable as Dahl seems to suppose. One example is provided by Smith (1990a) in his analysis of the development of the agricultural policy community in Britain. A set of historical and institutional procedures developed which effectively screened the agricultural agenda from the purview of all organized groups bar the National Farmers' Union (NFU). Another example is provided by Ryan (1978) where he describes how the Radical Alternatives to Prison (RAP) group was given less access to the centre of decision-making at the Home Office than the Howard League for Penal Reform. However these pictures can be made more amenable to pluralist analysis. Despite Smith's arguments that the agenda was 'controlled' by the NFU and the Ministry of Agriculture, Food and Fisheries (MAFF) he provides little evidence that other groups tried to change that agenda. It is all very well to argue that the reasons put forward at the time for protection of British farmers were not in fact very good reasons – for example, protection did not really aid Britain's balance of trade, yet Treasury support was secured – but whether this should have been recognized at the time by other groups, say British consumers, is another question. Where conflict did arise over the policy it derived largely from various elements within the government, and the NFU mobilized itself to fight changes with tactics similar to those of other well-organized pressure groups (Smith, 1990a, 120–23). Other groups who did fight the agenda, such as environmentalists, were simply not very well organized until comparatively recently. Indeed, when the policy community was threatened by health and environmental issues, highlighted recently by salmonella in eggs and Bovine Spongiform Encephalopathy in British cows, other groups have mobilized to break up the closed community, as Smith (1990b) has shown elsewhere. Remember, in pluralist group theory groups only mobilize when they see their interests threatened. Smith does not demonstrate that groups did perceive their interests to be threatened by the agricultural community or that they should have done so. Similarly, with Ryan's example of RAP, Jordan and Richardson (1987a, 190) argue that it

> is difficult to see how RAP could be welcomed to 'frequent' contacts, as was the Howard League, when they set out to abolish prisons and disapproved of liberal reforms on the ground that 'reforms simply reinforced the system'.

Others studying the organized group process argue that distinctions need to be made between 'legitimized' and 'non-legitimized' (Kogan, 1975, 75) or 'insiders', 'outsiders' and 'thresholders' (May and Nugent, 1982; see also Grant, 1978; Benewick, Berki and Parekh, 1973; Finer, 1966; and, for a similar distinction with regard to local government, Dearlove, 1973; Newton, 1976). These distinctions all lead to the conclusion that some organizations have more power than others, some are more readily heard than others, some are on the inside track and some on the outside track. But pluralists need not be concerned by these revelations; they are all contained in early pluralist works which do not argue, as many have maintained, that power is equally spread through the community but merely that all legitimate groups have access and may create costs for bureaucrats and politicians. Their power is a function of the intensity of preference, size of membership and degree of organization.

However, the pluralist case, though itself based upon analytic group theory, must not be so trivialized as at times appears to be the case, even with modern pluralists. Jordan and Richardson, influential British pluralists, write of the 'logic of negotiation' and provide an impressive array of evidence for the consultative process in Britain (Richardson and Jordan 1979; Jordan and Richardson 1987a, 1987b; and bibliographies therein), but this process must not be misunderstood. The 'logic of negotiation' and the 'consultative process' are key terms which can only be understood in the light of some of the examples they give in the terms of their essentially 'systems' (Easton, 1965; Deutsch, 1966) understanding of the modern state. In the 'systems' approach to the understanding of the state, each element contributes in some way to the whole. Parties, pressure groups, citizens and every other political object have a function which creates inputs into the political system. These are then processed into political outputs or policies. A balance, stability or equilibrium is maintained, for if state policies do not satisfy demands, then further demands are made which will be processed to maintain stability. Jordan and Richardson rarely refer directly to Easton, Almond or other systems theorists (but see Jordan and Richardson, 1987a, 5–6) but the systems language which has permeated political science (Dowding and Kimber, 1987) comes more directly to the surface in their work.

The heart of the concept of 'consultation' is contained in a passage from Henderson often quoted by Jordan and Richardson. Henderson argues that officials 'protect' themselves

> by making sure that, at every stage of the policy process, the right chairs have been warmed at the right committee tables by the appropriate institutions, everything possible has been done and no one could possibly be blamed if things go wrong. (Henderson, 1977, 189, cited Jordan and Richardson, 1982, 93, 1987a, 176)

Here 'consultation' consists of no more than warming chairs – bringing people in so that it looks as though everything possible has been done – rather than actually taking any notice of the people sitting in those chairs. Jordan and Richardson (1982, 86 original emphasis) argue that this form of consultation is part of the 'logic of negotiation':

> There is a *functional logic* to consultation and negotiation. Consultation contributes to system maintenance not only because it imparts a sense of involvement, but also because it should produce more acceptable policies.[1]

Consultation and negotiation, then, serve one major function: to maintain the system. They give the appearance of outcome power to some groups, which keeps them in order and adds legitimacy. However, as a by-product of that functional logic, groups do have an impact upon policy outcomes. They could be said to have some outcome power. Critics of pluralism can agree with the functional part of the description but not with the implied by-product. Nordlinger (1981) argues that policies only broadly reflect group aims when those aims fit in with what state actors wanted all along. Concessions to groups are only peripheral, and do not truly deflect state preferences. That is no form of outcome power for the consulted group. I do not think that Jordan and Richardson will disagree with this characterization, but they still draw a different model of the policy process. This is because they never actually address the central question at issue between the competing models of the policy process – who has power? (One of the few times they use the term 'power' is to state that 'it is wrong to see the unions as totally devoid of power under the Conservative governments after 1979' (1987a, 183).) This is not an oversight. After arguing that the power debate has run out of steam, they say (1987a, 5): 'when they find the questions impossible to resolve, political scientists have found other questions'. However, studiously avoiding important questions does not resolve them and sometimes leads analysts down strange paths. Jordan and Richardson's extraordinary pluralist analysis of the decision to buy Trident can only be attributed to their forgetting that different models of the policy process are essentially about who has power, when and why. It deserves to be quoted at some length:

> the decision to purchase Trident can hardly be described as involving a policy community of groups. However, whilst the policy is not made with the integrated participation of group spokesmen, the decision *is* made in terms of affected interests. For example, on Trident there is the interest of the research bureaucracy at Aldermaston, there is a wish to aid British Shipbuilders, a job-creation aspect. There are arguments in favour from part of the military hierarchy – and arguments against from the parts of the forces 'squeezed' by the demands of the nuclear budget. This is not to suggest that the critical decision to build Trident was the outcome of pressure-group politics – merely that with Trident-type decisions, there is a group dimension. Thus groups do not have insider access to influence all kinds of policy – but even where policy is evolved internally, in the longer term it will only be tenable if it can be 'sold' to an influential constituency. (Jordan and Richardson, 1987a, 179–80, original emphasis)

The tortured nature of this attempt to make the decision to buy Trident – something which was kept from all but a few members of the Cabinet – into an example of pluralism beggars belief and ought to turn the passage into a classic for critics of rhetoric. Notice the first sentence. The secret decision to buy Trident 'can hardly be described as involving a policy community' – surely a trivial truth – but the decision '*is* made in terms of affected groups' – another trivial truth: for all policies affect people, whether made by elected governments, faceless bureaucrats or absolute dictators. But what is meant by this statement is that governments will only carry out policies they think they can get away with, and if they attempt otherwise that policy will be overturned. If it is not overturned then, by definition, the policy was 'tenable' and 'sold to an influential constituency'. Either way pluralist group theory is verified for, again, it remains true no matter who makes the decision – whether an elected government, a faceless bureaucrat or an (almost) absolute dictator. If pluralism can really be taken this far then its critics are truly wasting their time, for all decisions in all polities are pluralist. All that is required is that policies affect people (and, as Brian Clough might say, if they do not then they should not be on the agenda) and that governments think that they can carry them out without causing a revolution. Finally, note that, according to this extraordinary passage, policy is not 'made', or 'decided' internally, thereby apparently giving some state group power, but rather 'evolves' amoeba-like from the slime of the internal policy process, thereby suggesting that no particular state actors have power. Whilst I have suggested that analytic interest group theory is trivially true, if pluralism corresponds too closely to group theory it too will be trivial. But pluralism must be an empirical not merely an analytic theory.

The policy community literature is an interesting way of organizing pluralist thought but it cannot be expected to give explanations of policy outcomes. Rather it is a descriptive theory which leads us to ask the questions pertinent to policy outcomes. Describing some policy community as 'open' and another as 'closed' tells us about the nature of the network of group and state actors. These different networks explain why one community tends to produce one sort of outcomes whilst another produces another sort; but this description does not explain why these different sorts of policy community exist (*pace* Rhodes, 1990). This requires a higher-level theory of preference formation, interest recognition and group mobilization. It requires, in other words, a theory of power.

6.3 The autonomy of the state

One of the more recent challenges to the prevailing pluralist and indeed elitist pictures of power at both local and national levels is that which argues that the state has relative autonomy both from the groups described in traditional pluralism and from the wealthy individuals and capitalist groups of elite theory.

Skocpol (1985, 9) writes:

> States conceived as organizations claiming control over
> territories and people may formulate and pursue goals that

are not simply reflective of the demands or interests of social groups, classes, or society. This is usually what is meant by 'state autonomy'.

The term 'simply' is important here. The state may pursue objectives which reflect to some degree the interests and demands of groups in society but not 'simply' those demands. The state has brought its own interests into the equation too. At times this claim is insipid. Where one person would describe public order as in the national or the public interest another may describe it as in the interests of the state. Whilst these different claims obviously have different normative implications there is no obvious empirical difference. The claim that the state brings in its own unique interests is not always unempirical, but in order to recognize a disjunction between the interests of the state and the members of society we need some means of recognizing state interests. By the methods of Chapter 3 such interest recognition requires a method of recognizing beliefs and desires or, failing that, of belief and desire assumption. How can the state have beliefs and desires? Only through those conscious beings who fulfil state roles. A claim about social outcomes which relies purely upon structural features pertaining to the organization of the state and society is not a thesis about state autonomy – for it is not a thesis about autonomy at all.[2] In order to make sense of the autonomy, or relative autonomy, of the state thesis we need to examine more closely the nature of autonomy.

We have seen that an actor is a causal agent: someone who brings about outcomes which would not otherwise have occurred. Actors are autonomous to the extent that they act through their own self-will; they are not controlled by anyone or anything. We explain action by the elements of belief and desire. Desire motivates and belief channels the action in a certain direction in order to satisfy the agent's desires. If action is explained outside these reasons, then the agent is not autonomous. Within this explanation autonomy means first of all self-will or freedom from the intentional constraint of others. It is in this sense negative freedom. It also usually denotes freedom from constraints which may control an individual 'behind her back', so to speak. These are constraints generated by the external world which are not intended by anyone and perhaps not recognized by anyone as constraints. If they were random then we could not proclaim them to end agent autonomy; they are only autonomy-reducing when they systematically alter actions in a particular direction. If any and every influence outside the individual were thought to be autonomy-reducing, then there could be no autonomy; for each and every individual is constantly assailed by influences from the world around them.

Further, why should we think that an individual living in a vacuum and therefore uninfluenced is more autonomous than those who have access to forms of knowledge other than the purely *a priori*? Moreover, the vacuum itself influences the individual afflicted by such a habitat. Therefore, if there is such a thing as autonomy, random influences cannot be thought to be autonomy-reducing; they just influence autonomous decisions. Neither can a person who receives all logically possible information be proclaimed 'absolutely autonomous', for she would be likely to behave differently if she received only a fraction of that information, or if she received the same information in a different order. There is no such thing as absolute autonomy, only relative autonomy, in much the same way as there is no such thing as absolute motion, only relative motion. All actors' actions are based upon their beliefs and desires, which are at least partly engendered by other aspects of the universe around them. Actors are only autonomous to the extent that they bring something to their actions themselves. This something I assume all people have.

The fact that all autonomy is relative does not, of course, entail that we cannot speak of degrees of autonomy any more than we cannot speak of degrees of motion despite all motion's being relative. The degree to which an agent is autonomous is the degree to which that agent is influenced in the relevant (i.e. non-random) manner by the world around her. This boils down to who influences whom and by how much; and to what influences whom and by how much. The first is the measure of others' power over the agent. The second is the degree to which agents' interests and beliefs are engendered by their social location. Chapters 4–6 attempt to elucidate the first, Chapters 3 and 7 the second. The degree to which actors are influenced by others and the way in which they are influenced depends upon two sets of general factors: (i) their 'internal resources': personality or character, including the degree of their self-interest, the degree of their risk-aversity, the amount of self-respect they have, their strength of will, determination, and so on; and (ii) their 'external resources', which largely depend upon their social location.

Of course, the two sets of general factors interrelate. How successful one is will affect the degree of one's risk-aversity, strength of will, self-confidence and so on. Given one's personality traits, however, social location will structurally suggest one's own interests. Thus we would expect individuals in one group to behave in certain sorts of ways in order to secure their own interests.

Modern group theory is based upon the assumption that we can make hypotheses about individual interests based upon social locations. If a set of individuals share some common characteristics then those characteristics will suggest some common interests. We can thus base explanations of the individual behaviour of masses of people on the groups to which those individuals belong. We have seen that we have to be careful in our explanations of group behaviour, for individuals will not necessarily pursue group or common interests because of the 'freerider' problem. However, with this problem allowed for, we can develop a pluralist picture of society. This much-criticized picture does need to take into account all of the groups in society, and the different sets and types of resources that all of these groups enjoy. We can see that some groups are privileged in that they have access to important resources which other groups do not enjoy. This is the basis of Nordlinger's argument that the state must be brought into the equation in a way that pluralist theory does not allow.

Eric Nordlinger (1981, 1) asks two questions at the beginning of his *On the Autonomy of the Democratic State*:

> How can we account for the authoritative actions of the democratic state, its public policies broadly conceived? To what extent is the democratic state an autonomous entity, one that translates its own policy preferences into authoritative actions?

The first question is one which this book tries, in broad terms, to answer, as any which is concerned with political power should do. The second question, from a social choice perspective, is nonsensical even under Nordlinger's elucidation of the notion of the policy preferences of the state. Arrow's theorem demonstrates that no voting rule can be guaranteed to aggregate a disparate set of preference functions into a single social choice function satisfying the minimum conditions of rationality specified in Chapter 2 (Arrow, 1963; Mackay, 1980; Riker, 1982). Nordlinger (1981, 11) defines the state as

> all those individuals who occupy offices that authorize them, and them alone, to make and apply decisions that are binding upon any and all segments of society

and he explains the preferences of the state as 'public officials taken all together' (Nordlinger 1981, 3), which is itself explicated as

> the resource-weighted parallelogram (or resultant) of the public officials' preferences. ... State preferences, as distinct from the discrete preferences of public officials, are those with the weightiest support of public officials behind them, based on the number of officials on different sides of the issue, the formal powers of their offices, their hierarchical and strategic

positioning relative to the issue at hand, and the information, expertise, and interpersonal skills at their disposal. (1981, 15)

Nordlinger recognizes that, whilst the term 'state preferences' suggests a unity, it should not be taken as such, for many officials will not have preferences on many issues and they will not all be in agreement on all issues. He also understands that the preferences of state officials can be further explicated in individualist terms based upon personal preferences suggested to them by their social location. In this way he hopes to avoid the charge of reifying the state (though denying reification is not the same as not reifying). It is clear that if Nordlinger does not reify the state then he only avoids doing so in so far as he resurrects individualist group theory by acknowledging that there are groups of state actors who share common interests based upon their location within the state structure.

Rational choice models have been offered to explain the behaviour of different state actors. Downs's (1957) early model of the party process explains the behaviour of politicians as they strive to be elected and to form society in the way in which they desire. Political parties are coalitions of divergent groups which push and pull policy, but the final direction is structurally suggested by the constraints of election. In order to be elected politicians must appeal to many divergent groups in society, and so will make their promises as vague as they can in order to make as many groups as possible believe that their wishes will be carried out. Politicians will make different appeals to different groups, a process made easier by modern methods of identifying and contacting voters. Politicians will try to make other parties become more specific and will advertise the policies of their opponents more clearly than they will specify their own. Once in power, politicians and governments will try to set the political agenda and try to shift the whole ideological spectrum more towards their preferred view (Dunleavy and Ward, 1981). If readers believe that such deliberate ploys smack of crude conspiracy theories, the example of Alfred Sherman and John Hoskyns, sometime advisers to Mrs Thatcher, is instructive. Sherman and Hoskyns believed not only that perceptions had to be changed, but at times wrote policy suggestions deliberately so radical as to be laughable. Their intention was for such ideas to shift the centre ground of the political spectrum in their preferred direction. Later their previously mocked ideas were to become government policy (Young, 1989).

It is not only through argument and propaganda that governments try to shift the political spectrum, but through the policies themselves. Many aspects of the policies of Thatcher's government in Britain can be seen as shifting the basic consensus. As a result of pushing the sale of state-owned council housing to private ownership and encouraging small-time shareholders and small business, sector-based voting allegiances have been shifted towards the right. Of course, such strategies are not confined to the right: Morrison's famous pledge to 'build the Tories out of London' through council housing was essentially the same strategy.

Rational choice models of other state actors abound. Niskanen (1971) argues that bureaucrats will try to maximize their budgets, for then their own income and self-esteem may be enhanced. In so maximizing the state interferes in more and more aspects of citizen's lives. Bureaucrats have advantages over their political masters through their near monopoly of information and may thus control the direction of even the most radical governments. The problem for Niskanen-style models is explaining how radical governments have managed to cut the budgets of state bureaucracies through privatization and tighter control (Dunleavy, 1986).

Patrick Dunleavy's (1985, 1989) more sophisticated and empirically more accurate models provide solutions to these problems. He argues that the idea of budget maximization is far too simplistic. Inflating the overall budget of a bureaucracy is not in the interests of bureaucrats. Dunleavy argues that there are several types of budget within bureaucracies. The *programme* budget consists of all the expenditure the bureaucracy controls even though parts of this budget may be passed on to other public-sector agencies. The *bureau* budget is that part of the programme budget which is directly implemented by the bureau. The *core* budget is that part of the bureau budget which is spent on maintaining its own operations: that is, the money spent on its own staff, their accommodation and administration. He also identifies the *super-programme* budget (Dunleavy, 1989) which is monies spent by other bureaux over which the higher bureaucracy has some influence. Dunleavy (1989) distinguishes eight different types of bureau depending on the functions they serve. The relative sizes of the four budgets vary across the bureaux. The difference in bureau functions and the variable relationship between the different budgets and officials' welfare levels also varies across the eight categories. Where policy is administered by large line-bureaucracies there is a close connection but where policy is administered by systems of bureaux the connection is weak. The interests of senior bureaucrats in all agencies are promoted by bureau shaping rather than maximizing. Senior bureaucrats seek to shape their bureaux into small, elite and centrally placed staff agencies rather than routinized, troublesome and conflict-prone line-agencies. Privatization has thus provided an opportunity for senior bureaucrats rather than an attack upon their interests.

A similar self-interested form of privatization may be seen in the housing departments of many local authorities in Britain. Faced with threats to housing departments from decreasing budgets and the possibility of private landlords taking over housing stock, some councils have looked to the possibility of setting up non-profit-making housing trusts staffed by the personnel of their housing departments. These suggestions have often been taken up enthusiastically by that personnel, and why not? It will secure their jobs and give them greater freedom from the interference of elected officials, especially over such questions as their working environments and salary scales.

In this way simple models of 'state autonomy' become more complex and realistic models of 'state actor autonomy'. At times these state actors work in harmony, at other times they work at variance. Bureaucrats and elected politicians share some interests but diverge over others. Local and central state actors have divergent as well as convergent interests, as do departments within the central administration. Explicit recognition of the autonomy of state actors rather than 'the state' explains otherwise paradoxical sentences contained in the first part of Smith's (1990a, 219–20) argument that

> measures that create autonomy may also limit it. In creating the agricultural policy community, the state was expanding its capabilities in order to develop an agricultural policy community. In doing so it was reducing its autonomy. The establishment of the community removed decision making on agriculture from the Treasury, the Foreign Office and the Board of Trade and gave it to the Ministry of Agriculture. So the policy community gave the state the ability to develop an agricultural policy but it removed the ability to change that policy in the future.

If we cut out the reification of the state in the quotation we see that what Smith means is that, by concentrating power in the hands of the bureaucrats at the Ministry of Agriculture, the government made it harder for itself or future governments to change the policies through its other agents at, particularly, the Treasury – the usual bastion of anti-protectionism and anti-subsidization of the British economy. Thus, non-reification of the state means that we should concentrate upon individual and group actors within different segments of the state, for these segments have divergent interests as well as common ones. However, if these individualist arguments, and indeed Arrow's theorem, apply to the idea of the state's lacking a single preference function, why do they not apply to group theory too? They do. The 'freerider' problem is just a special case of the intransitivity of group preferences based upon individual preferences. The simple concept underlying group theory is that we can recognize the individual actions of people in group terms. It gives us an entry point into the complex explanation of mass behaviour. A group is defined as a number of people who share a common interest. Quite where that 'interest' occurs within their complete preference function is not specified but we can assume it is an interest in the way 'interest' was described in Chapter 3. This interest gives rise to a 'proper' group (Truman, 1951, 24) when the common interest leads individuals to interact and behave politically. We can explain why one group behaves in one way, and another group in another way, even though some individuals will be members of both groups and thus sometimes behave in one way and sometimes in the other. Group theory can thus encompass Nordlinger's analysis if we break down his notion of 'state preferences' into smaller group units: the interests of national bureaucrats in different types of bureau, nationally elected politicians, local bureaucrats, locally elected politicians, the police, social workers, teachers, and so on. These disparate groups of state actors take on specific roles and affect policies in specific ways according to their common group interests and their powers based upon their resources, including the state resources identified by Nordlinger.

In Truman's analysis, and in the work of later group theorists, the government and the bureaucracy themselves constitute interest groups. The importance of Nordlinger's work is that it reminds us of the special resources of these privileged groups of state actors. Whilst this is acknowledged in early group theory and early pluralist thought, it has been largely ignored in modern analyses of the behaviour of groups in society. These tend to concentrate upon organizations representing the interests of groups of citizens. There are, to my knowledge, no examples of overtly pluralist or group theoretic studies of unorganized groups, and the government or the bureaucracy is usually represented as the actor in the middle of the 'parallelogram of forces'.

In what sense can there be said to be an autonomy of the state? Nordlinger suggests that there are three possible forms of state autonomy based upon the autonomy of state actors:

I strong form: when state actors act on their own preferences where these clearly diverge from those of society;
II medium form: when state actors actively change the preferences of society;
III weak form: when state actors' preferences do not diverge from those of 'society' – here, though, it is just as reasonable to assume that state actors are acting on their own preferences as on those of 'society'.

This can be expressed as a two by two matrix (Figure 6.1).

Figure 6.1: Explanations for the authoritative actions of the democratic state

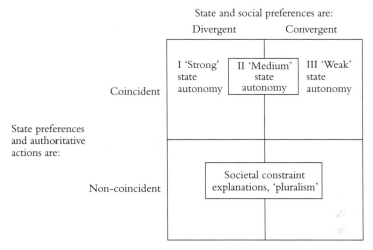

Source: Modified from Nordlinger, 1981, p. 28

Nordlinger claims that, whilst we may be able to identify states which by and large correspond to each of the three models, the models may each be applicable to any given state in different issue-areas. Thus the bureaucracy at the Ministry of Education may act in the weak form, at the Treasury in the medium form, and at the Home Office in the strong form. Elected officials, both guiding and being guided by those non-elected officials, likewise take the weak, medium and strong forms as they move from ministry to ministry.

Nordlinger gives a series of autonomy-enhancing strategies (47 in total) which state actors may pursue in order to promote their own interests. They reduce to a number of categories:

(a) employing the *capital resources* of the state;
(b) using its *authority*, when mediating conflict in a biased way (ignoring some groups, only pretending to consult and so on);
(c) utilizing its *control of information* to help shape other actors' preferences (it may hide its real actions, obfuscate decisions, carry out legislation but underfund programmes, and so on);
(d) *agenda management.*

Category (b) corresponds to Harsanyi's second category, category (c) to Harsanyi's first category (Chapter 4, page 69). Agenda-management is another form of information control or manipulation, whilst the state's capital resources may be used unconditionally to affect outcomes or conditionally in the form of a threat or an offer to provide incentives to other groups. Nordlinger explains the autonomy of elected officials as being generated by (i) their lack of real need to remain in politics in order to secure their own material interests, and (ii) the fact that in the main they have secure constituency pluralities, so they can ignore the electoral process. Both reasons apply to US politics more than to western Europe. In western Europe the much stronger party system itself creates a constraint upon elected officials, though the higher one rises in the party bureaucracy the higher one's power over the party itself. When in government the party machine tends to be driven by the elected government, rather than the other way round; when out of power the relationship is more variable. There are other major constraints on elected officials, which depend upon the cycle of regular or frequent elections. They must take account of the wishes of the electorate in a very broad sense. Indeed we may see a significant constraint upon the actions of elected governments with regard to major economic objectives: which constitute a major determinant of voting behaviour. One aspect of this constraint has been studied extensively by Adam Przeworski (1986). These arguments lead to structural accounts of power and the structural challenge to pluralism.

6.4 Structural accounts of power in society

Przeworski argues that the capitalist democratic system, with its regular cycle of elections, constrains any democratic socialist party from changing the economy from a capitalist path of development to a socialist one. Przeworski's model assumes that (1) a socialist economy is in the material interests of the working classes, (2) the working classes are in an overall majority, (3) all classes vote for the party which is in their own material interests, (4) the shift from a capitalist path of development to a socialist one will bring about dislocational costs, and (5) the time frame of these costs is longer than the electoral cycle. From these assumptions Przeworski shows that the electoral cycle under capitalist democracy may constrain socialist parties from bringing about socialism. We can see this in Figure 6.2.

Here s–s' is the material well-being of the majority working class along the socialist path of development, c–c' the material well-being of the majority working class along the capitalist path of development. The working class would obviously prefer to be on the socialist path. However, if they are on the capitalist path at time t_1 when society tries to move towards a socialist path, their material well-being will suffer, reaching

Figure 6.2: Capitalism to socialism across the valley of transition

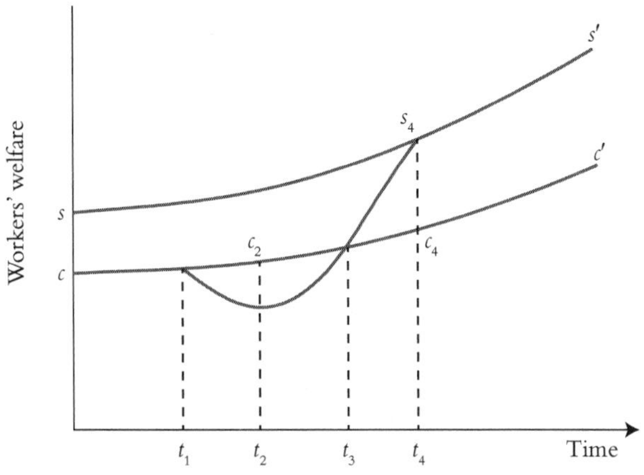

Source: Modified from Przeworski, 1986, p. 177

its lowest point at t_2. This is much lower than if they had stayed on the capitalist path (at c_2), regaining the same level as the capitalist path at t_3, and reaching the socialist path at point t_4. Although the point which the workers reach at t_4 by leaving the capitalist path (s_4) is higher than it would have been had they remained on the capitalist path (c_4), there is a valley that must be traversed if they are to get there. Why would workers, led by a socialist party, not be willing to traverse that valley? We may answer that question by utilizing a Downsian model of the democratic process.

Downs (1957) assumes that political parties are coalitions of individuals who share certain views about policy. As such parties do not have single preference functions, but rather their official policies are compromises between their different factions, compromises which are largely driven by electoral considerations. The electorate's perception of how competent each party is to run the capitalist economy is largely based upon how well it actually runs when each party is in power. If moving from the capitalist to the socialist path of development does require traversing the valley, then this course will only appeal to those who will gain in the long run (by assumption the majority working class); and only to that group if they can see the benefits of such a transformation. However, if we imagine that the time between t_1 and t_4 is, say, 30 years but the electoral cycle is only four to five years, we can see that a democratic socialist party is in some difficulty. It would have to persuade the working class of the long-term advantages of their socialist policies over five or six election periods. Przeworski suggests that this is not possible.

I mention Przeworski's model here not because I think it represents an accurate picture of the difficulties facing democratic socialist parties – though it may do (see King and Wickham-Jones, 1990, for a critique of the model in relation to Britain) – but for two reasons: first as an example of structuralist argument which is developed from individualist and rational choice assumptions; second, as an example of the sort of constraints that may operate upon elected state actors. Certainly social democratic parties in power have faced a flight of capital which has affected their policies. Harold Wilson's account of the 1964–70 Labour government reveals continuing worries about the reaction of foreign investors to Labour's policy (for example, Wilson, 1974, 58–65, 323–5); and how a popular speech can make a difference, as apparently did Wilson's to the Economic Club in New York (Wilson, 1974, 135). (Though Wilson's worries over major union strikes take up far more space in these memoirs.) Capital, being controlled by relatively few capitalists, does not require collective organization in order to constrain government, whereas labour does (Offe and Wiesenthal, 1985). For example, a story – 'Labour win "will cause City panic"' – in the *Sunday Correspondent*, 29 April 1990, suggests that money invested in Britain will be moved overseas if Labour were to win the next election. The story emanated from interviews with 31 City fund managers controlling 60 per cent of City funds. Seventeen of the managers interviewed stated that they would reduce the share of money invested in British equities if Labour were to win. Whether their statements are true or whether they are merely trying to affect the result of the election, the power of these managers seems clear. But there is no question here that we are discussing individual power. Further the costs of capital flight are relatively low, certainly lower than the costs of shifting to new forms of capital accumulation. Capital flight is thus individually rational and collectively very powerful. Hence the autonomy of elected state actors is restricted by certain requirements of the capitalist system. Nordlinger's strong state-autonomy model is most likely to occur only in policy sectors where the bureau has a clear conception of its role and strong political backing, particularly where business is weak and divided and not trusted by state actors (Atkinson and Coleman, 1989, 52).

Capital constraints are systematic in the sense that they are non-random; they depend upon a relatively enduring set of relations between individuals denoted by the functional roles they perform within a society. The system of relations is such that no one single actor is necessary or sufficient for the continuance of the system, and if one or more individuals fail to fulfil their roles there are strong incentives for other actors to take their place.

The constraints upon state action exist because of the type of economy under which the elected state officials act. State officials' actions are modified by economic factors because they face re-election. This view is the 'more plausible' view attributed to Marx by Jon Elster (1988, 208), where the interests of the capitalist class serve as a constraint upon government rather than as a goal for its actions. Elster (1988, 213) claims that this does not give power to capitalists:

> it is tempting to argue that if the choice between the feasible political alternatives is always made according to the interest of one group, then it has all power concentrated in its hands. On reflection, however, it is clear that power – real, as opposed to formal – must also include the ability to define the set of alternatives, to set constraints on what is feasible.

Elster imagines a situation where the government wishes to maximize tax revenue, which is a function of the rate of taxation. If tax rate is 0 then revenue is 0; if tax rate is 100 then revenue is 0, for no taxable activity will occur. The actual position of the Laffer curve (Wanniski, 1978) will determine the optimal rate of taxation and where it lies is an empirical matter over which capitalists and government alike have no control. If it is high then capitalists are unlucky; if it is low they are lucky. Capitalists may be just lucky that what is in the interests of the government is also, by and large, in their interests too. They have no need to intervene. Elster also imagines that the power of both government and capitalists is limited by a number of factors. They are in a bargaining game, where capitalists may not need to intervene – partly because they are lucky and partly because government will not act in ways which are too contrary to their interests in case they may be provoked to intervene, an example of the law of anticipated reactions. We must not forget from our bargaining model of power that there are always costs of intervention which stop groups intervening too readily, a fact which many power studies forget (Ward, 1979, 217). This can still imply an autonomous state in the sense that the state actors do not act as they do because it is in the interests of capitalists, but because it is in their own interests. However, as *it happens*, it is in their interests to do what is in the interests of capitalists, although capitalists have no control over this fact. It is an empirical fact beyond their control. This argument does not demonstrate that the state always acts in the interests of capitalists, but that, even if it does, it does not follow that the state is not autonomous nor that capitalists have any great power. They may be merely lucky.

However, it may be premature to say that capitalists are *merely* lucky. First, collectively they have some control over where the Laffer curve lies, though not enough to say that individually they have power. They do not actually have to act collectively in order to utilize that collective power, for they collectively have that power even though they act individually. However, acting collectively may utilize that power most effectively. There is little evidence that capitalists have organized so as collectively to utilize that power. Second, and more importantly, whilst capitalists may be lucky, this luck may be systematic. If it is systematic it is still not power, but it is more than *mere* luck. Their luck is systematic because it attaches to certain locations within the institutional and social structure. Luck here is non-random but rather may be predicted methodically. However, this systematic luck is still luck under Barry's technical definition, for individuals get what they want without trying. They do not use their resources in order to achieve the desired outcomes. Systematic luck is not the same as power for, though it attaches to individuals in certain positions in society and it attaches to them *because* they hold those positions in society, they individually have no control over those outcomes. Individually they are not powerful. Collectively they may have some control, but if they do not act collectively they are not getting the desired outcomes through their efforts. These outcomes are achieved through the efforts of others who are working for their own interests. In the Przeworski model capitalist interests are looked after by the government for the interests of the party and the politicians which make up the government. Thus collectively they may have power, but they are *actually* getting the desired outcomes through luck. Collectively they are powerful as well as lucky. Often getting what you want may be the result of the law of anticipated reactions. This is often described as a form of power. However, if you get what you want without trying, because of the law of anticipated reactions, then you are lucky. If, however, this operates effectively as a result of the reputation one has managed to produce for oneself – as in the favourite mafia example – then one has produced reactions through an exercise of power. For here one's reputation has been deliberately engendered, and we have seen how important reputation is in modern bargaining theory (Chapter 4, pages 73–77). The distinction is a fine one, but one which needs to be made in order to break the verbal deadlock between pluralists, elitists and statists.

Nor is systematic luck a form of 'structural power', though it may well be what those who use that term mean. Chapter 1 provided two arguments against the use of the term 'power' in conjunction with the term 'structure'. First, it is redundant, and second, it is a misuse of the term. The relationships between people do indeed help to constitute the incentive structure facing actors but in so far as power is a causal notion they are not the cause. As argued in Chapter 2, structures are the background conditions, not the cause of outcomes. A cause is an event, and structures are not events; actions are. 'Systematic luck' may be an oxymoron, but it is not a nonsensical phrase like 'structural power'.[3]

6.5 Business and the policy process

Although capitalists may not need to intervene in the policy process under liberal democracy in order to ensure governments that pursue policies broadly in their interests, it does not follow that they do not so intervene. Nor does it follow that it is not in their interests to intervene. The state may pursue policies broadly in my interests, but that is not as good for me as the state's pursuing policies which are in my particular interests. Business does intervene in the policy process, and not only is it an important pressure group, it is also a privileged one, both for the Przeworski-style structural reasons and also for other sorts of reasons. These other reasons include both the close relationship between capital and government through their institutional links and the close relationship through their personal links. There is much evidence, particularly from America, that people from similar social backgrounds take up important bureaucratic state roles (in both local communities and national government) and within business (Mills, 1956; Domhoff, 1978). I will argue that close links foster the outcome power of businessmen and capitalists but that this is eased by the structural relationship demonstrated by Przeworski which brings luck, not power, to business and capital.

We can see from the collective action problem that it may well be to the overall detriment of capital for businesses to pressurize governments to act in the individual interests of different companies or industries. A favourite argument around the time of the 'ungovernability' hypotheses of the 1970s (Brittan, 1975; Olson, 1982; Birch, 1984), although far more work was concentrated upon the labour side of the equation, there is much evidence to show that this also applies to the business side (Ingham, 1984; Grant, 1987). We may ask why, though the ungovernability theories clearly apply to *both* capital and labour organization (Brittan, 1975; Olson, 1982, 41–7), so much work went into the union side of the equation and relatively little work on the capital side. One answer is that political scientists are ideologically biased against trade unions and support the capitalist system. That answer is barely worth the mention. A second answer is that the ideological hegemony of the capitalist state leads us to study the labour side and not the capitalist side. Perhaps this is the case, but this answer looks to me more like a redescription of the question than an answer to it. A third answer, which may perhaps be worked up to something which puts some flesh onto the second, is that studying the union side is easy whereas studying the business side is hard. It is harder for us to study the power of business precisely because it is a privileged interest group. Its outcome power is more concealed than overt trade union activity and so we are less likely to plan studies of that power and the costs (in all senses) of the studies we do plan are relatively higher. (Furthermore, there is more openness in trade unions that amongst businessmen. The culture of secrecy of the latter is probably a result of competitive pressure where it pays to be close-mouthed.)

Beyond the study of the business pressure lobby (or 'capital') as a single group, one must recognize the conflict between different elements of the interest group 'capital'. The interests of finance capital are often at variance with those of productive capital (Ingham, 1984); and within the latter category small businesses often have very different needs from big businesses (Grant, 1987). Also, of course, competition between firms within industries and between industries gives rise to sectoral cleavages in the clamour for government attention. In this section I want to put aside the structural resources which give rise to capital's privileged position (a side which also underlies the analyses of Lindblom, 1977; Marsh, 1983; Marsh and Locksley, 1983; and Grant, 1987) and concentrate upon those non-structural factors which privilege it.

The distinction between finance capital and productive capital has been recognized as very important, particularly by British commentators who view it as part of Britain's supposed decline (Coates, 1984). However, the distinction between the two is not as clear-cut as many have believed, and certainly the comparison with other nations does not set Britain as far apart as is sometimes imagined (the following is based upon Grant, 1987, ch. 4). Productive firms have tied up many of their monetary assets in financial deals which have little to do with their productive business and have become even more involved in finance as 'takeover fever' has taken over. The development of the professional corporate treasurers within large firms is one aspect of this convergence. Directors of companies transcend the finance/production divide as multiple directorships multiply, whilst financial companies began joining the Confederation of British Industry (CBI) from the late 1970s. However, this interlocking of the two should not blind us to arguments advocating the dominance of finance capital over productive capital. Grant (1987, 79) is correct when he writes 'financiers and industrialists seem to recognize that what unites them is more important than what divides them'. What unites them is making money; what divides them is how they go about that process. That British government policy has for a long time served the needs of finance capital rather than productive capital is almost undisputed; that this has led productive capital to become more involved with finance as a result should be of no surprise and should not lead us to think that the interests of the two are becoming closer. What is happening is that finance is taking over from production. In the present environment it is the fitter genotype.

It is difficult to verify why successive British governments should serve the interests of finance capital so well. Certainly, the structural argument is important, but that argument can be used to serve productive capital equally as well as finance capital. It cannot be just that it is easier to make (and lose) money in the financial markets, for that is a part of what is to be explained, though such an argument might be worth pursuing. But there are a number of particular explanations which may work and are not necessarily in competition with *each* other, though they are difficult to quantify in importance. Michael Moran (1983) argues that the City has been able to convince state actors that its broad interests were beyond the power of State control and were thus not on the political agenda. This control was possible because of the close links the City enjoys with the Bank of England and the Treasury. These links are not merely in personnel but, perhaps more importantly, in the type of economists who serve these institutions. It is a commonplace that commonplace economists find finance easier to understand than productive industry; it fits better their mathematical and econometric models. Furthermore, it may well be, as Ingham (1984) argues, that the practices and power of the Treasury and the Bank of England are better served through practices serving finance rather than productive capital. The oft-cited example of this precedence of financial interest is Labour's reluctance to devalue the pound from 1964 until devaluation was forced upon them in 1967 (see Blank, 1978; Coates, 1984; Grant, 1987). Again, we must not overestimate the lobby power of financiers, for they do not always get what they want. The Labour government of the period did not reduce public expenditure in the way financiers would have liked, whilst the Selective Employment Tax was a specific measure designed to reverse the trend of employment shifting from productive to service sectors (Graham, 1972). This suggests that the luck of individual financiers depends upon the collective power of international finance making them privileged in the Przeworski sense. This is more important to securing their general interests than their lobbying efforts or any ideological hegemony (Stones, 1988). The latter is not unimportant but is hard to gauge. The more diffuse 'ideological' power of the importance of finance capital to the economy within the minds of the electorate, for example, is seen nightly on the British television news when the FT 100 index is read out and we are told whether it has gone up or down a few points – something of no earthly significance even to those who hold a number of shares.

The complaints of the elite and state theorists is that traditional pluralist studies concentrated upon *governmental* decision-making and took little account of the decision-making which takes place outside government, in both private and quasi-public bodies. This is no longer true of modern pluralism which has taken on board some of these views. (Or rather it is no longer officially true of modern pluralism. Pluralists still often concentrate upon government decision-making, but admit of non-government decisions within policy communities.) Modern pluralism talks of policy communities and policy networks, comprised of sets of groups, organizations private and public around which decisions are made and policies determined (for example, Heclo, 1978). This substantial advance from the more undifferentiated approach of early group theory should enable us to see how power becomes concentrated and certain groups left out. However, we should not lose sight of the power issues altogether, as some pluralist analyses have done. Rather the extent of elite and state actor control of the policy issues, together with the impotence of unorganized groups, should be the central questions to be asked within any given policy area. That requires studying not only the overt social power of the actors concerned but also the unintended checks upon political action. Models of the structure of different policy arenas are the first stage of analysis, the formation of those structures the second.

6.6 Conclusions

Who, then, has power? Everyone has some power, but some have more power than everyone else. The 'some power' that everyone has is outcome power, and for most people that independent outcome power is very restricted in scope. The outcome power to which individuals may contribute in collective organizations is less restricted but is still constrained by economic considerations. Education, housing, environmental and racial policies may all be affected by the pressures of well-organized groups but the impetus to organize is the major obstacle to gaining that power. The poorer, less articulate, less interactive and larger groups face greater collective action problems than richer, more cohesive and articulate and smaller groups. The nature of development within communities may also be affected by group activity, but whether or not development should take place is less open to influence. Economic questions are also less open to influence and the other policies are less affected when they become subsumed under the economic issue, which is increasingly the case in the modern welfare state. The social power that some have over others is similarly based upon resources they may use to change the incentives for other actors to behave in ways they otherwise would not have. But we can begin to see that the types and degrees of power that individuals and groups in society have depend upon the various sets of resources that they bring to the bargaining table we call society. The sets of resources reduce into the categories identified by Harsanyi: (1) information, (2) legitimate authority, (3) unconditional incentives to alter others' incentive structure, (4) conditional incentives to alter others' incentive structure. Two other elements are also important: stubbornness and the freerider problem.

Information

Social power is, remember, the ability to affect the incentive structures of others in order to bring about, or help bring about, outcomes. One of the most subtle ways of affecting others' incentive structures is the sort of information one feeds to them. In that sense we are all in the hands of the mass media, though how massed they are against our interests is a moot point. For group analysis the power of organizations often depends upon their relationship to state agents. For example, one of the reasons why the British NFU has been so powerful in keeping the agricultural policy community closed is the close relationship it has enjoyed with MAFF and the fact that it is the main supplier of factual data for the Ministry.

Information is a key empirical dispute between pluralist models of the policy process and statist ones. Pluralists concentrate upon the information supplied by pressure groups to the public and to state agencies; statists concentrate upon the information which is then utilized by state agencies in policy formation or, in other words, the information passed by bureaucrats to politicians and by both of them to the public. In order to resolve the respective power of pressure groups and state groups we must take into account the degree to which these two-way flows of information are causally efficacious in policy formation and implementation.

Legitimate authority

In the final instance policy is decided by state actors, principally the government. This is an important aspect of state power, but power based upon legitimacy does not end here. Within pluralist theory influential groups are those organized groups which are recognized as legitimate. Organizations which represent constituencies outside what is genuinely acceptable to society are not a part of the policy-making arena. For Dahl (1956, 1986) in the United States that includes groups representing communist beliefs and, as we saw, in Britain that includes such organizations as Radical Alternatives to Prison, whose demands are too extreme for society to contemplate. Or rather, it might be objected, whose demands are too radical for the state to contemplate. Again, a key empirical dispute between pluralists and statists is how far these two different interpretations of the evidence are sustainable. How far do state and elite actors decide which views are beyond the policy pale and how far do they decide this because of the signals proffered by society? Certainly, many radical pressure groups start their campaigns directed at the public and only turn their attention to state actors after they can point to wider societal support for their views. Resolving such disputes between pluralists and their critics is not easy. Governments do refuse to take account of certain organizations because they claim they do not legitimately represent the interests of those they are supposed to represent; but does this entail Nordlinger's weak-autonomy model where state actors only follow those groups whose preferences fit their own? Certainly, as Nordlinger argues, the evidence is just as compatible with his model as any other. However, it is equally true that there are many examples of changing public attitudes leading subsequently to changing government policy, which have been originally caused by organized pressure group activity. Again, there is a dialectical relationship between pressure groups and state groups. At times they rely upon each other, but they are often opposed. Organizations require resources beyond themselves, such as public opinion or party support in order to push against unwilling state actors. However, it would be wrong to see this mobilization process as something above and beyond pluralist theory.

Unconditional incentives to alter others' incentive structure

Increasing the costs or benefits of alternative courses of action for other groups is a major power of groups in society. From the state side any changes in the legal situation constitute such unconditional incentives. From the capital side also such unconditional incentive shifts occur as development takes place in one sector of the economy or one geographical region of the nation rather than another. This is a, if not the, major wielding of social power but one which is usually missed by the case-study approach of traditional empirical pluralism. However it is also a difficult process for power-holders to control and deliberate exercise of power should not be equated with all shifts of incentives within society. A deliberate change of economic policy which creates unemployment in order to reduce the bargaining power of trade unions is a very different situation from a slide to unemployment which has the same, but unintended effect.

Conditional incentives to alter others' incentive structure

This is the most obvious form of bargaining power, for it requires the actions of one group to be recognized and responded to by another. It is a form of power in which the participants have to be aware of what is taking place. The resources of the groups engaged in such bargaining are easier to quantify, for the groups themselves must make those resources perceivable as a part of the bargaining process. Of course, the power of groups goes beyond their capital and human resources and in overt bargaining includes the skill of the negotiators themselves. There is a sense with this form of power that one's actual power may be measured by the benefits one receives. However, even with this overt form of power, one (or one's opponents) may not have used one's resources in the best possible manner, so the actual benefits may not necessarily track one's power in all possible worlds.

Stubbornness

Everyone can be stubborn. Therefore everyone can stop others pushing them around. The degree to which people are stubborn depends upon the way in which they make the calculations of long-term utility and the nature of their personality. This latter factor is outside my brief but is obviously important to personal power relations. However, the costs of being stubborn may be immense – perhaps too great to contemplate for most people in many situations. There was little to gain from stubbornness for the miners of Clear Fork Valley where their homes and employment prospects were threatened if they remained on strike (Gaventa, 1980, see Chapter 5). The costs of stubbornness rise when one has others to worry about as well as oneself – which is why Dr Stockman thought that 'the strongest man in the world is the man who stands alone' (Ibsen, 1960, 104). However even the power of stubbornness is only the power of stopping others having overt social power over one, for it does not enable one to have any outcome power at all (except in the trivial sense of stopping certain outcomes where one is pushed around), nor does it necessarily stop others having more subtle forms of social power.

Checks upon group power

The most important block upon group power does not come directly from the social power of other groups. The mobilization process is itself the most important block upon group power. The resources of individuals within groups are important to that first mobilization, and those resources include not only relative wealth, energy and individual skills but also information about the world by which to recognize one's own interests. The internal constraints upon mobilization are themselves a product of others' outcome power but not necessarily an intentional product of that power. This major check upon group power is all too often lost in the major pluralist works, which concentrate upon the activities of organized groups, which have already overcome their first major hurdle. States differentially respond to mobilization as different models of the policy process suggest. This area of group power is perhaps the least subjected to close empirical analysis and yet is the one which affords the best test of the elite, state and pluralist analyses.

The next chapter will be mostly concerned with ideology – something which is often supposed to be a tool of the powerful to keep the weak in their place. However, ideology may also be used to help the weak to overcome collective action problems. The discussion of the development of preferences by actors in different social locations contained in Chapter 7 is intended to add to our methods for discovering the powers of unorganized as much as organized groups.

Preference Formation, Social Location and Ideology

7.1 Introduction

Ideology is every bit as controversial a topic as political power, and one where it is probably even harder to reach definite conclusions (see Hamilton, 1987, for a review of the many different ways in which the term has been defined). However, a book on political power cannot ignore the subject altogether. Indeed, the notion of ideology has entered into the discussion in earlier chapters in various forms, but something a little more systematic will be attempted here. Clearly, I cannot do the subject full justice in one short chapter of a short book.[1] I do not intend to provide a complete account of ideology but rather to explain how an ideology helps the power of some groups in society. Here the term 'ideology' is used not in its overarching sense, but rather in the sense of 'the trade union ideology', 'the culture of the Home Office' or 'the Treasury world view', and so on. Thus I will not attempt to explain ideology *simpliciter*, or 'the ideology of capitalist society', or 'the dominant ideology', or whatever it might be – though I do believe that such a project would follow the lines I sketch in this more limited version. The more limited sense is important in a consideration of power in society, for two related reasons. First, it helps to explain preference formation as well as to describe the preferences of actors. Second, analysing the generation of such ideologies helps to explain some actors' power when the ideology is deliberately created or encouraged and used by those actors, and to explain actors' luck, including systematic luck, when it develops naturally or unintentionally.

In Chapter 3 I discussed various elements of preference schedules and their relationship towards individual interests. I also discussed very briefly some elements of preference formation and the nature of group interests. I have also referred in many places to the concept of 'systematic luck', defining it in Chapter 6 (page 132) but promising to explicate it in greater detail in this chapter. I will try to explain how individual interests are, in part, created by individuals' social locations and relate that to the discussion in Chapters 1 and 2 of the nature of the social explanation offered in this book and by rational choice theory more generally. This then leads to an explication of systematic luck which will to some extent close the gap between Brian Barry's concept of luck and 'Personal Identity Luck' to which it was contrasted in Chapter 4. Finally, the relationship between preference formation, power and ideology will be discussed.

7.2 Preference formation and social location

Individuals act as they do because of their beliefs and desires, which also give rise to their preference orderings. A complete account of why individuals think and act as they do must therefore analyse how individual beliefs and desires are generated. Desires are the hardest to explain of the trilogy belief, desire and action. It is difficult to show why particular individuals have the particular desires they do. It is a task for psychologists or perhaps sociologists. Instead rational choice theory makes certain general assumptions about desires. It assumes utility maximization and generally (though not always explicitly) assumes that a monetary value can be assigned to that utility. It also generally assumes self-interest, though that concept is broadened to include one's close family and even friends. It rarely goes beyond that, however, though altruism has been shown to be a concept which can be brought into the domain of rational choice theory.

I have suggested a simple maxim of sociological explanation – simple beliefs and desires require simple explanation, whilst complex beliefs and desires require complex explanation. The idea behind this is that we cannot really *explain* why someone prefers strawberries to raspberries; we can only note that indeed they do. If someone switches from a preference of strawberries over raspberries to raspberries over strawberries we may examine why this has occurred. Perhaps they have eaten an excess of strawberries over the previous weeks, or perhaps they had a nasty experience whilst eating strawberries and their disquiet over the event has transferred to the taste of strawberries. The latter is a more complex explanation than the former, but the evidence may point in that direction. However, if we could offer no *reasoned* explanation of the preference switch – one day A said he preferred strawberries and the next day raspberries, and his behaviour confirmed this – then we cannot explain it. This need not worry the social scientist. Some people prefer raspberries to strawberries and some people strawberries to raspberries and that is that; and in this case A preferred one on one day, the other the next. This does not cause undue problems for rational choice theory, even though it may seem to break the condition of transitivity. Rational choice can encompass people's changing their minds over time (one could not simultaneously prefer one fruit to the other and the other to the one), as long as they do not do it too often or too frequently. (Changing your mind over which fruits one prefers by the minute would be puzzling and require complex explanation.) Simple desires thus enter our models of the world as an assumption or a description of what people actually prefer as demonstrated by empirical analysis.

Complex desires, however, do require explanation, partly because they resemble beliefs as much as simple desires. If George expresses the desire to marry Anne or prefers the idea of marrying Anne to that of marrying Mary, he does not have the leeway to state the next day that he has changed his mind in the same way that he might over preferences in fruit. Partly we might think that this is because marriage is a serious business, whereas eating one fruit rather than another is not. If a child were to change his mind suddenly over an expressed desire to marry one person rather than another, we would not give it a second thought because we would not take the matter seriously. But this is not the only, nor the most important reason. The main reason why this is unacceptable is that we cannot understand how a person could honestly change such a desire unless there were *reasons* for it. One does not need to have reasons to have the taste-buds one has, but one does to have reasons for desires over an issue which involves a serious commitment and the complex web of conventions, norms, and so on which are contained in the very idea of marriage. This is because one has to have an understanding of the concept of marriage in order to be able to express a desire to be married or to prefer to marry one person rather than another. If we feel someone is incapable of having that complex set of beliefs then we do not take seriously their idea of marrying someone – as in the case of a child – and so we can accept without demur any change in their preferences on the matter without real explanations for it. This is why such complex desires are more like beliefs than simple desires, for they *involve* a whole set of beliefs. The sorts of propositions in which individuals express desires and which concern the power analyst are complex in this manner. Individual interests are complex, involving both simple and complex desires and beliefs. With the assumption of utility maximization and certain assumptions about simple desires, the rational choice theorist may explain why individuals at different social locations have different sets of interests. Given the model of preference formation which underlies this analysis, she may also explain how individuals come to have preferences which seem at variance with their interests – as we saw in Chapter 3. We may also see how individuals come to accept beliefs which may be termed 'ideological'. In part this may occur because we impute a set of beliefs to the individual more complex than that individual attributes to himself.

In Chapter 3 a distinction was made between 'endogenous interests' and 'exogenous interests'. Endogenous interests are those we have by virtue of our simple desires or by virtue of decisions we make regarding features of the world. Exogenous interests are ones we have by virtue of the circumstances in which we find ourselves. Thus a coal-miner has an interest in the wages and work conditions of miners, a shop assistant in the wages and work conditions of shop assistants. These are their exogenous interests. Their attitudes towards nuclear power would be counted as an endogenous interest since it is not so dependent upon their particular circumstances. However, a coal-miner might be worried about nuclear power in terms of its effects upon future need for coal. The distinction is not always an easy one to make; hence it was described as an ostensive categorical distinction and not one that is supposed to divide the world into natural categories.

Understanding exogenous interests can help us to understand why people form different sets of preferences, given their social locations, and to arbitrate between the two sides in the debate between those who want to argue that individuals have free will and those who argue for greater determinism. I assume that everyone has free will – it was one of the bases of the 'stubbornness objection' to a determinist bargaining model of power in society – but also recognize that each person is influenced by the world around in two related ways: by a situation effect and by a perspective effect (Boudon, 1989). The former leads people to see the world in a certain way because of their social situation. The second gives them a certain sort of perspective on the world as a result of their interests, which in turn depend upon their social location. This is why I prefer the term 'structural suggestion' to 'structural determination'. In this way the interests which are exogenous to one rational choice model describing a given outcome are endogenous to another rational choice model which explains how those interests arose. The latter type has a simple explanatory schema (Roemer, 1986b). We may define two elements which enter the explanation of any given outcome: the institutions, property relations, resources and technology at any given time t, R_t; and the list of all people and their preferences at time t, P_t. Roemer suggests that there are two ways of combining R_t and P_t to explain two different sets of outcomes. First, there is the 'solution process' which gives rise to the production and distribution of goods, new institutions and so on: $\{R_t, P_t\} \rightarrow R_{t+1}$. Thus R_t, P_t gives a causal explanation of the institutions, property relations, and so on (R) in the time period $t+1$. However, we can also explain (P) using the same two elements $\{P_t, R_t\} \rightarrow P_{t+1}$:

> Thus individuals are formed by society, and these individuals react rationally to their environment to produce tomorrow's environment, which in turn produces individuals who think somewhat differently from before, and react in their environment to bring about yet a new equilibrium. (Roemer, 1986b, 196)[2]

As Roemer rightly recognizes, if preferences are entirely a product of the actual conditions at prior time periods, then preference could be left out of historical explanation as long as we have a conversion manual to turn R_{tn} into R_{tn+1}. As he also says, though, in doing so we would miss the explanation of why history had unfolded as it did. However, a model of preference formation does not require such strong determinism and can allow individuals to bring elements into the process of forming their own preferences. Indeed, to the extent that we discover our preferences by the very act of choice, there may well be a random or chance element in preference formation also. As Roemer suggests, a person may, in some sense, 'choose' the preference ordering he wishes from the environment he inhabits. However there

> will be expectations upon a person by the rest of society, given the place he occupies in R_t, which will direct him to order his options in certain ways. A person born into the Rockefeller family will maximize his satisfaction by choosing to remain a capitalist, by and large. It is more accurate to say culture chooses a person's preferences for him. But culture can be understood as ideology, and if there are rational foundations for ideology ... then the process of endogenous preference formation can be seen as a rational choice. The social formation of the individual can be explained whilst at the same time requiring that society be understood as the consequence of many individuals' action. (Roemer, 1986b, 199)

Section 7.4 will look at the rational foundations of ideology.

7.3 Luck and systematic luck

It seems that, whilst the institutions and social relations of society are formed by the actions of individuals, those same structures suggest the preferences and interests of the acting individuals. The structure provides incentives to act one way or another and the best way of maximizing individual utility will be suggested by one's social location and the possibilities for action which suggest themselves. In Chapter 3's definition of objective interests I provided an account consistent with free will, and sensitive to the sensible aspect of privileged access, but which also explains how individuals may mis-specify their own best interests despite being able correctly to specify their preferences. Whatever one needs to fulfil one's desires (all things considered) is in one's interests, even if one does not recognize this.

It is easy to state objective interests in this way, but it does not answer hard questions. Przeworski's model of the impossibility of socialist democratic parties' moving to the socialist path of development from the capitalist assumes that it is in workers' interests to do so. For the purposes of such a negative argument, the assumption is perfectly justified. But it says nothing about the real interests of a given worker. Even if workers are better off in the long run with such a dislocation of the economy over, say, a 30–year period, it does not follow that such a dislocation is in my interests, even if I am a worker. I might be dead in 30 years' time and not care about future generations. We cannot assume away the latter clause and retain a liberal attitude towards desires (unless we assume all desires are endogenously determined and go down the determinist path I want to avoid). As long as we assume some free will, then those basic, simple desires of each individual come into the calculation of objective interests, even though, as I demonstrated with regard to such simple desires (the bread example of Chapter 3, page 33), they may well be based upon social structure. However, what we can do is to explain how it is rational to take on certain interests given one's position in society, and how this may lead to systematic luck. Some groups of people are lucky: they get what they want from society without having to act. Some groups are systematically lucky: they get what they want without having to act because of the way in which society is structured. As noted earlier, 'systematic luck' is an oxymoron – if I always win because of the way the world is structured, then in what sense can I be said to be lucky? Despite the apparent contradiction I think the term 'systematic luck' is well chosen. It denotes the property that some individuals have of getting what they want without trying – Barry's luck – but also the fact that this property attaches to certain locations within the institutional and social structure – it is systematic. This luck (or fortune) is closer to 'destiny' than to simple chance. Winning all the time without trying is not the same as wielding power; winning all the time without trying when one's efforts cannot affect the outcome is not the same as having power. We must recognize the disjunction between getting what one wants and outcome power, even when we expect people with certain properties to get what they want and, in the main, they do. We can explain this by recalling that in rational choice models of social situations actors are denoted by their structural properties and not by their individual or personal properties. They are 'capitalists', or 'bankers', or 'bureaucrats', or 'junior Home Office officials', or 'citizens', or whatever it might be. Actors in this structurally denoted sense have certain powers based upon the resources they have as those actors. Capitalists have capital to lend: that is what makes them capitalists, and that is a resource they may use. Bureaucrats have institutional resources to affect the information flow to politicians and the public, and other authoritative resources to affect the relationship between groups and government. Bureaucrats may use these as a source of power. In the same way as these actors have resources from being who they are, they may also be lucky as a result of who they are. Actors in this structurally denoted sense have a certain amount of luck which is derived from the institutional and social relationships they bear to one another.

We can give several examples of this luck, based upon the structure of relations between actors. The first is the Przeworski model. Here the interests of capitalists are secured by competitive parties pursuing economic policies to maintain a strong capitalist economy in the interests of the electorate. Capitalists do not need to intervene and may not have the power to intervene but their interests are secured. By getting what they want without trying they are lucky in the Barry sense. But this luck is not entirely contingent: they are lucky because they are capitalists in a capitalist system with a competitive party structure. They are systematically lucky because their luck is dependent upon that system of relations. They may be powerful as well, but there is an empirical difference between the two. If they are systematically lucky and not powerful, then if their interests are challenged they will not be able to respond; if they are also powerful then they can respond. Similarly, property-owners are lucky in the Growth Machine model of community power studies. Developers, politicians and the majority of the local citizens may be in favour of new development from which property-owners gain relatively more than everyone else. The property-owners may gain without trying, and where they are unable to encourage growth on their own they are lucky. Such luck is systematic in this case to the extent that others' interests are not entirely contingently correlated with growth in a property-owning society. A third example, which I will develop a little more fully, is the luck of being a farmer in Britain this century. It is an example which may show (depending upon developments over the coming years) that farmers in Britain were lucky without having much power and what power they did have was based upon their earlier luck.

The standard line on the power of the farming lobby in Britain has maintained that their power is based upon the well-resourced, -financed and -organized lobby of the National Farmers' Union (NFU). This organization has maintained close links with the state in a corporatist or neo-corporatist fashion through the Ministry of Agriculture, Food and Fisheries (MAFF).[3] Both industry and agriculture in Britain have well-resourced interest groups, but agriculture has enjoyed a high level of support by comparison with industry which has a relatively low level of state intervention. There are a number of reasons for this. Some relate to the resources of the organizations representing agriculture and industry. In the agricultural case the NFU has had a virtual monopoly of representation over the past 50 years, whereas there has not been such a monolithic representation of industry. Further, the interests of farmers are also more homogenous than those of industry. These are not the only factors which have affected the respective state handouts. Smith (1990a) (see also Self and Storing, 1962) has shown that the power of the agriculture lobby developed for specific and partly contingent reasons just prior to and during the Second World War. A prewar consensus developed to maintain agricultural support for farmers given (a) the problems confronting farmers in the face of the falling price of their produce through the operation of the free market, and (b) the need of the nation to feed itself during wartime. This support was institutionally secured when the War Cabinet decided in 1942 not only to maintain agriculture during the war but to promise support after it. This promise was made in part in order to secure the confidence of farmers who had felt badly let down after the First World War when subsidies had been withdrawn between sowing and harvest in 1921. The postwar dollar crisis also helped the farmers' cause, since the Treasury decided that support for agriculture would help to ease matters. Smith's account does indeed reveal an outcome-powerful NFU in the postwar period, but it also reveals that that power base was founded upon luck: first, the luck of war, which increased the importance of agricultural production over the production of many other goods; second, the luck of the dollar crisis, which the Treasury (probably incorrectly) calculated would be eased by supporting agriculture. In both instances the farmers' power was secured by factors outside their control. This is not to deny that the farmers did lobby government and that the specific support they secured depended upon the signals they sent. But it is to deny that the help they received was determined only by those signals. They were lucky in the sense they would have got something like what they wanted without trying. Their outcome power was to mould that 'something' into what they specifically desired.

Farmers have systematic luck too. They have the luck of just being farmers. Food and food production are always perceived to be of central importance to the nation. This type of luck closes the gap between Barry Luck and Personal Identity Luck somewhat, and in this sense farmers – in comparison, say, to small businessmen – are systematically lucky the world over. They may get what they want without trying. The basis of this systematic luck is also a power base, for the fact of their (perceived) greater importance may be used as a bargaining device to secure governmental outputs more to their liking. This distinction is fine, but it is one worth making: in the first case the farmers get what they want without trying, and in the second they have to act. The difference is therefore empirical as well as a finely tuned theoretical one, and the fact of groups acting or not in pursuit of their objectives is one of the bones of contention between pluralists, elitists and statists. Recognizing the distinction between social power, outcome power, luck and systematic luck lets us see that simple observation of group behaviour without the theory of individual and collective action does not allow us to pass judgement on that debate.

The NFU was lucky in a fourth way too. In the 1930s and 1940s it did not face a set of competing interest lobbies. Whilst there were environmental concerns, the farmers were perceived to be the guardian of the countryside, not its enemy. (One of the few contentious issues in the immediate postwar period was rights of way over farmland. Legally the farmers largely lost the argument.) However, the past 40 years of intensive farming have seen this guardian image change. Farmers are no longer perceived as friends of the environment but destroyers, whilst the healthy food lobby has become increasingly important as food scandals increase in number and profile (Smith, 1990b). In part this is because these issues have appeared on the agenda because of the nature of modern farming, and lobbies have formed as individuals see their interests threatened. This seems to be good pluralist theory, but Smith (1990a) argues that before the 1980s other lobbying organizations were 'excluded'. However, with the sole exception of the example of the conscious decision to exclude Stanley Evans and the Farmers' Union of Wales (Smith, 1990a, 144–5), little evidence of exclusion is offered. Rather:

> The 1947 Agriculture Act which laid the foundation to [sic] post-war policy was very close to the proposals of the Royal

> Society and the NFU. That there were shared values on the future of agricultural policy made the creation of a policy community that much easier. That, MAF, the farmers and the Government agreed meant that they could work together on the establishment of agricultural policy. (Smith, 1990a, 99)

True, other interests were not organized even where those interests were beginning to be recognized and without their knowledge being acted against. True, other badly organized groups were not consulted as they might have been had the issues been more contentious. But, as we have seen, neither of these truths should lead us to take on board Lukes's multi-dimensional power, nor to introduce the confused concept of structural power, both of which Smith does. The farmers had some power and they had some luck. The degree to which they have either may be assessed by seeing how the agricultural policy community changed when these contrary groups organized.

The changing policy style – Smith (1990b) describes it as a change from a policy community to an issue network – is dependent upon the emergence of new issues and lobbies. The importance for the luck/power argument, however, is that it provides a good test of the power of the NFU. Previously the farmers got pretty much what they wanted without action and with few contrary pressures to face; now they face a multitude. How powerful the NFU really is in the face of these changing demands remains to be seen. The counteractual to be considered is how it could have coped in the 1950s, had those demands been made then. Smith's (1990a) argument, which places most power in the hands of state actors at the Ministry of Agriculture and the Treasury, suggests that the NFU would not have been able to sustain its power to any greater extent than now.

7.4 Ideology

Rational choice theorists have always seen ideology as a cost-saving device (see, for example, Downs, 1957, 98). Rather than developing a completely specified viewpoint on each issue, individuals relate issues to their general philosophy or viewpoint, thus saving themselves time and energy. Ideology in this sense is a satisficing device for limiting individuals' feasible sets of beliefs. For example, why does some individual who moves to the Home Office suddenly take on the Home Office view? There are two reasons. First, underlying these sorts of viewpoints are the interests of the people located within these institutions. There is no contradiction between a worker's believing that trade unions are generally too militant and that the nation suffers as a result and his voting for strike action when it comes to his own trade union dispute. This is simply a group-level collective action problem involving all workers. Similarly, there is no puzzle in C. Northcote Parkinson's apocryphal story of the official who fought for funding he had opposed after moving from the Treasury to a spending ministry. In both departments we may say he was merely doing his job.

Second, even where the ideology is not in the overall or long-run interests of the group the individual has joined, it may be rational for him to take on that view rather than attempt to change it. The official may readily take on the view of the Home Office because it is cheaper for him to do so than to challenge that view. The Home Office line on any particular issue has already been drawn and all he has to do is take it from the top drawer and read it out. To challenge the ideology would involve working out a complete and coherent set of answers to any problem which arises. This is apart from the possible effect upon his promotion prospects of challenging the official ideology. Again, even though a new viewpoint may actually be in the interest of everyone within the group, the group faces an individual-level collective action problem in developing a *new* viewpoint.

The power of groups may be increased by these sorts of ideologies. If new members absorb the group's views, it is easier to overcome individual-level collective action problems. Rather than needing to bargain within coalitions to decide on a course of action, the individual within the group may just read off from the ideology how to respond to threats or initiate action. For organizations representing functional interest groups this allows much readier response and potentially increases the outcome power of the group as a whole. Such ideologies, however, also reduce the outcome power of individuals within the group to change policy, for the status quo policy always starts from an advantageous position. The advantage of the status quo is well known from both casual observation and formal social choice theory (Riker, 1982).

Where a group manages to extend its viewpoint beyond its members to those within the policy-making community or beyond, then obviously ideology is an important weapon within its armoury. Within the terms of the bargaining model of power, the group ideology is captured by the information variable. Groups which wield their control of information for their own purposes rely upon the rationality of ignorance over many matters. There are four elements of rational ignorance (Goodin, 1980, 38): (1) people have imperfect information, (2) they know they have imperfect information, (3) it is costly (a) to acquire more information and (b) to assess more information, and (4) the expected gains from further information are thought likely to be less than these costs. Only when it becomes apparent that one's interests are suffering does it become worth while to gain more information. With something like the food industry that threat to interests will become apparent first to those who work on health issues but are not a part of the food industry itself and thus have no reason to hide that information. Once you have two sources of information it may become worth while to try to assess the information received. Where information is deliberately withheld or tampered with then we have overt manipulation. Goodin argues that manipulators play on one or more of the elements of rational ignorance by four distinct strategies: first, lying (1); second, secrecy (1, 2 and 3a); third, propaganda (1 and 4); fourth, overload (3b). Of course, there are costs to these strategies. First, one faces penalties if caught; and second, one faces a credibility problem in the future, with consequent reduction of influence.

Extending a group viewpoint beyond the group to the policy community and to the nation or world as a whole takes us into the murkier waters of ideology in general or dominant ideologies. There is some dispute about whether or not ideological belief-sets must be false in some overarching sense and whether true beliefs can be ideological. There is less dispute that ideological beliefs must work in the interests of some groups and to the detriment of others. Any actual belief-set is bound to include some false beliefs; those sets, or at least the false beliefs contained within them, may be said to be ideological, particularly if those beliefs work in the interests of some groups and to the detriment of others. Where true beliefs work in the interests of some groups and against those of others this suggests that the former groups are lucky. Personally I would shy away from calling true beliefs ideological even if they work against the interests of some groups, though I have no philosophical argument so to discount them. Rather I agree with Boudon's (1989, 205) observation at the end of his rational choice account of ideology:

> despite the crisis of values which we hear so much about nowadays, one value in particular remains unchanging and certain, so much so that we might say that it is independent of all historical and social conditioning, and that in this sense it can be regarded as transcendent. This value finds expression in the fact that most people unconditionally prefer the truth to its opposite.

If true beliefs work against one's preferences then one is unlucky; here we must conclude that sometimes, for some people, life is tough. Where false beliefs work in the interests of some and against those of others we do seem to be in the heartland of ideology. A rather more complete rational choice account of ideology, which brings in both truth conditions and interests, may be developed.

One of the best-known rational choice accounts of ideology is that of Jon Elster (1983a, ch. 4; 1985, ch. 8). Elster has a strange concept of individual belief which leads him to develop a psychological view of ideology where a structural view would be more appropriate (though Elster describes his view as a structural one). Elster's psychological account of ideology stems from his specifying the way in which false beliefs develop through the growth of irrationality caused by poor but prevalent reasoning and people's psychological propensities, which he has been studying for so long. Ideological beliefs for Elster derive from such distorting processes as cognitive dissonance, counterfinality, weakness of will, the sour grapes story, and others. However, whilst these psychological propensities may be utilized by manipulators to create ideology and may indeed aid its creation without manipulators, it is a mistake, I think, to equate the generation of ideology with these processes. Even without them someone may suffer from holding certain beliefs which are against their interests and which may be rightly described as ideological. Further it does not seem unreasonable to think that those beliefs which are caused by cognitive dissonance or counterfinality may not always be ideological ones. The root of my disagreement with Elster derives from his weird view of beliefs. He writes (1985, 462):

> I shall define the ideological in structural not functional terms – as an entity, not a certain type of effect that one entity may have upon another. Broadly speaking ... these entities are beliefs and values consciously entertained by some individual or individuals. They are entities, that is, which (i) exist, (ii) exist in the minds of individuals and (iii) exist consciously for these individuals.

Certainly, if there is such a thing as ideology then ideological beliefs exist. But I do not think we can say that they exist 'in the minds of individuals', nor that they are necessarily conscious. The beliefs which we have undoubtedly exist in some sense,[4] but not necessarily *in the mind* – beliefs are not the same as thoughts which pass fleetingly through the mind, but rather are relatively enduring. An individual's beliefs at any given moment are those propositions to which she would assent if she were asked, and those propositions which are entailed by the first set of propositions. I will give an example of the first condition, and then some examples of the second, much more controversial condition. Even though it is open to debate the second condition is important for developing an account of ideology.

If I am asked, 'how do I get to Old Marston?', I will answer the question differently according to where the questioner asks her question. If she is in Barton, I will explain that she should go up the hill to the Green Road roundabout, take the fourth exit onto the ring road, turn off at the first exit, bear right and then left, following the sign half a mile into Old Marston. If she were somewhere in London I would give rather different directions, but they would end by taking her up the A40 to the Green Road roundabout and taking the *third* exit down the ring road. What the route from Barton to Old Marston tells us is what I believe to be the relative directions from Barton to Old Marston by road. Now I have believed that to be the direction from the time when, if asked, I would have given those directions. I believed it before I had ever considered how to get from Barton to Old Marston but was able to give those directions (even if I had to think a moment before working it out). I believed that that was the direction to take well before those propositions were ever entities 'in my mind'. To take a simpler example, I believe that dipping my big toe in a duck-shaped glass bowl filled with strong hydrochloric acid will hurt. I have believed that ever since I knew both what my big toe and strong hydrochloric acid are, and that the shape of the bowl will make no difference to the effects of acid on human flesh. But I have never before this moment (honestly) considered the effects of putting my big toe in a duck-shaped bowl of strong hydrochloric acid. Beliefs are 'entities' under some understanding of that term, and they do exist, under some understanding of that term, but they do not exist (only) 'in the mind'. Nor do they exist consciously there. I am not at the present moment conscious of more than a few of my beliefs, and cannot be conscious of beliefs which I presently have, but have never yet, nor perhaps ever will, think about. But they are still my beliefs.

The second clause of the definition of belief given above is more controversial, but it is a necessary consequence of the first. The definition is: an individual's beliefs at any given moment are those propositions to which she would assent if she were asked, and those propositions which are entailed by those propositions to which she would assent if she were asked. The second clause follows from the first because of the meaning relationship of entailment. If I believe 'p' and 'p' entails the conjunction of 'r' and 's', then I believe 'r and s' even if I am not aware of the fact. Consider the following conversation:

> T: 'I consider that it's disgusting the way Maggie has brought
>   in the Poll Tax. Everyone's against it and I hope the Tories

> lose the next election because of it, and I voted for them last time.'

M: 'I'm not going to pay, I've not filled in any forms and if they catch up with me I'd go to gaol rather than have to pay.'

T: 'Well, you shouldn't do that. The Poll Tax is wrong and the Tories will suffer for that but you shouldn't cheat the taxman. I'll have to pay more next year 'cos of people like you. I suppose you were one of those who rioted the other week; people like you should be locked up; you make me sick.'

If we overheard this conversation we would be justified in asserting that T believes that laws passed by an elected government in the British Parliament are legitimate, even where he, and the majority of people, disagree with them. He further believes that Mrs Thatcher's government is legitimate, and that the British system of government is legitimate. We would be justified in asserting this even if T has no idea what the term 'legitimate' means. In an important sense, therefore, T holds a constitutional liberal ideology even though he is unaware of this, being ignorant of what liberalism and its attendant claims entail.

There is an obvious objection to the second clause. If I believe 'p' and 'p' entails the conjunction of 'r' and 's' then I believe 'r and s' even if I am not aware of the fact. But I may well consciously deny, say, 'r'. This seems to entail that I believe 'r and not-r' and from the law of non-contradiction it seems that I therefore believe everything. This objection is problematic, but I do not think it overwhelms the second clause. Certainly, in extensional contexts the damaging inference would ensue but within the intensional context of individual belief we may be more sanguine. It is true that where I hold contradictory beliefs the laws of logic mean that I *should* believe everything. However, if I do not want to believe everything then I must drive out from my belief-set some of the contradictory ones. This is the underlying reason why we try to expel beliefs from our belief-set when we realize that some contradict others. However, where 'r' follows from an explicitly held 'p' but I also explicitly hold 'not-C, I will drop either 'p' or 'not-r' unless I am not aware that the two contradict. The assumption underlying the second clause is that I am not aware of the contradiction because I do not know that 'p' entails 'r and s' and therefore do not attempt to drive out one of the contradictory beliefs, but continue to believe both. The fact that 'not-r' is held explicitly but 'r' is not does not allow us to say that therefore I 'really' believe 'not-r' and give that priority over 'r', since I also explicitly hold 'p' and that entails 'r'. Instead we have to turn to what the individual would assent to if the contradiction were to be pointed out. We can now see why the second clause is so important to an account of ideology, for it is only at this second conditional stage that some beliefs which are consciously held would be given up. It is the conscious holding of some beliefs which would be given up because of their contradiction with other beliefs about individual interests that has so captivated those who wish to develop ideological accounts of historical moments. The difficulties of understanding how we may come to hold contradictory beliefs due to 'perspective effects' and the complexity of the issues which appear intuitively simple are demonstrated in Boudon (1989).

From this account of individual belief emerge two elements in the development of an account of ideology which includes an interest and a truth component: first, that the ideological beliefs are false in some objective sense, and second, that the falsity is related to the way in which those false beliefs work for or against the interests of the individual. The account of ideology briefly offered here could be developed and related closely to the development of objective interests contained in Chapter 3. This would develop the way in which those entailments from explicitly held beliefs are not recognized despite contradicting other explicitly held interest-related beliefs, because of the way in which they are expressed. It would examine the way in which conventional implications of word and sentence become buried when they are moved from one context to another. However, this must wait for another occasion. In the bargaining model of power, these contradictory beliefs are propagated by the powerful under the 'information' variable or, if not explicitly propagated, those which gain from their currency are lucky. The relationship between ideology and power, therefore, is a close one and operates at two distinct levels. Ideology may be utilized by individuals or groups as a part of their resources in the bargaining game. It is also a source of luck, for it affects the way in which individuals view their own interests. In the less restricted sense particular ideologies help to bind collectivities together and help them overcome their collective action problems (Taylor, 1982, 1988). They act as labour-saving devices to enable individuals to identify their interests quickly whilst they get on with the business of living their own lives.

8

Conclusions

8.1 Why are we interested in political power?

There have been many books written on the subject of political power; there will be many more. Why is there this fascination to write and to read so many, so diverse accounts? One reason, of course, is that each writer believes all previous accounts of political power to be flawed and that advances may be made with a new approach. Presumably readers of such books believe the same. However, as we saw in Chapter 6, some empirical researchers have stated that, because of the problems associated with the concept of power, political scientists must find new and easier questions to answer. I do not find that adequate, for two reasons. First, scientists cannot avoid the hard questions just because they are hard. Secondly, and more importantly, I do not think it is possible to avoid the inequitable distribution of power in society. We may be able to avoid using the term 'power', we may be able to ask pettier questions about society, but ultimately any description of the policy process is a description about the structure of power in society. Necessarily so, for a description of the policy process shows how policies are formulated and put into practice. That involves uncovering the causes of the policy outputs; and explaining outcomes in society ultimately involves power, for ultimately it involves actors. What actors achieve and what they may and may not achieve unavoidably involves their power and its limits; even if we do not explicitly say so.

The first reason why we are interested in political power is that it is fundamental to our understanding of politics. Any description of politics involves the language of cause and effect, and where actors are involved it therefore uses the concept of power. Following Weber, I suggested that the very nature of politics involves conflict in the sense that without conflict there would not be any politics. But politics does not require that the conflict be overt or even recognized and thus much of human behaviour is potentially political. Another great German, Bismarck, remarked that politics is the art of the possible. Much of my study of power has involved studying 'the possible' and indeed much of the methodological dispute over power has been over different ways in which to analyse counteractuals – a subject to which social scientists have not devoted enough study. (Exceptions include Elster, 1978, and Morriss, 1987; and see works cited therein.) Power is intimately linked with the counteractual questions because it is a causal notion and analysing causation requires the analysis of what might have been as well as what occurred. One reason why power has been so contested over the years is that it is so difficult to analyse what might have been.

The analysis of political power is more problematic even than the analysis of causation, for power involves not just causation but also individual reasons for action. It involves individual wants and desires; people's beliefs, their preferences, their needs and their interests. I tried systematically to analyse these notions in Chapter 3. That analysis was empirical in the sense that I tried to show how we may analyse all these notions through careful study of human behaviour in conjunction with a simple and coherent theory of action. It was also normative, for it involved an important normative assumption: I assumed that individuals are the best judges of their own desires. I demonstrated that it does not follow from that assumption that individuals are always the best judges of their own interests. I also argued that because individuals may have desires for contradictory outcomes we may need to place individual desires in a sequential order within a preference schedule which may well go beyond an individual's own awareness. However the construction of such orderings requires careful empirical analysis which must involve basic assumptions. These assumptions can always be challenged, and it ought to go without saying that researchers are prone to error and so may mis-specify individual preferences and interests. What ought also to go without saying, though it unfortunately seems to require repeating *ad nauseam*, is that this epistemological fact about all researchers does not pollute the ontological fact about the possibility of individual error, indeed systematic error, about interests.

A normative assumption about individual desires is impossible to avoid. It is impossible to avoid because it is an assumption about what people care about. Morality is about values which are the currency by which we measure what people care about. The particular normative assumption I utilize is not incontestable but nor is it unjustifiable. I justify it by the claim (itself contestable) that individuals are actors who act for reasons and do so with their own free will. That is not to say, of course, that those reasons may not be examined. We may discover why people believe the propositions they believe, and desire those objects which they desire. We can discover the causes of those things, and even be critical about beliefs and desires because of the nature of the causes we discover. However, in the end, I assume that individuals do themselves bring something to the causes of their actions. I assume free will and not strict determinism. I cannot express this any better than Rousseau (1984, 87), who asserted that

> I see in all animals only an ingenious machine to which nature has given senses in order to keep itself in motion and protect itself. ... I see exactly the same things in the human machine, with this difference: that whilst nature alone activates everything in operations of a beast, man participates in his own actions

though I am less sure than Rousseau of the stark distinction between people and animals. This assumption of free will is just that: an assumption. Arguments may be given to justify it, but I will offer just one reason. If I did not assume it, I could not be bothered to study society or consider questions of moral and political philosophy. Since I am bothered, I assume free will.

The recognition of a basic normative assumption underlying my approach to political power reveals a second reason for being interested in the nature of political power. Normative questions surrounding the nature and distribution of power are just as important as, if parasitic upon, the empirical ones. They are parasitic upon them in the sense that the empirical questions require some resolution before the normative ones can even be specified. Political power is ineradicably a normative concept as well as an empirical one because morality, if it is anything, is concerned with how people should act and how they should be treated. Understanding agents' powers gives us an understanding of the limitations upon their possible actions, but also helps us to understand what they could, but do not actually, achieve. We all have the power to do things we should not do, or the power to do things we should do but do not. Describing the power structure in society reveals the constraints under which we act, and the way in which people are, and could be, treated. We are interested in political power because we are interested in the actual limits of what people can do, and in those things which people do which we feel they should not.

Both the need for counteractual analysis of political power and especially its ineradicably normative nature have led many analysts to claim that there can never be agreement over its definition. They claim it is an essentially contested concept (Lukes, 1974; Connolly, 1983). I would not have attempted to write this book if I felt the strong claim of essential contestability were true, though I am well aware that my arguments will not convince everyone that I am right. In Section 8.2 I will briefly provide some general arguments against essential contestability. Section 8.3 will then illustrate this argument by showing why I believe the rational choice approach provides the variously competing theories about the power structure with a method in which they may be evaluated one with another.

8.2 Contestation in political discourse

A very strict thesis of essential contestability takes the word 'essential' seriously. It states that if a word is 'essentially' contestable then there is something about the very nature of the word which entails that there can *never* be agreement over its meaning. This is a very strong claim. It does not merely state that as a matter of fact not everyone will agree over the correct application of the term, but that *necessarily* they will never agree. It is not merely an empirical claim, but a logical one.

It may be hard to make sense of essential contestability as a 'logical' claim. For agreeing or not agreeing are things that people do. Agreement is empirical in character, not logical. But the 'logical' interpretation can be weakened whilst remaining strict. We may suggest that people may perchance agree on a particular interpretation of a word but they can never do so *decisively*. Reasons may be given for various particular interpretations of a particular word, and one formulation may be accepted by all and sundry, yet the reasons for that particular interpretation do not logically rule out others. The arguments are not strictly deductive nor do the conclusions (the interpretation of the word) strictly follow from the premises. Thus there are reasons for one view or another but those reasons are not decisive. Connolly (1983, 225) suggests that saying something is *essentially* contestable is:

> to contend that the universal criteria of reason, as we can now understand them, do not *suffice* to settle these contests definitely.

He gives three claims of essential contestability: (1) that no previous or current philosophy has secured a basic set of concepts; (2) that future attempts will fail; and (3) that there are reasons why attempts to produce closure of debate will fail (Connolly, 1983, 229). The first two are compatible with the claim of mere contestation in political discourse and so do not need to be challenged. No one disputes that there has never been agreement, and that there are lots of reasons why there may never be agreement. The fact that we are all subject to the problem of knowledge is one. The interest of essential contestability comes with the third claim. It is interesting if it provides logical reasons for a lack of closure. If it entails the weaker claim that it is *unlikely* that full agreement will occur, then there is no reason for predicating the term 'essential' to that of contestability.

Non-decisiveness comes in two forms. The first occurs where there is no extant empirical evidence decisively to prove the superiority of one theory over another. This is sometimes explained as the open-textured nature of some concepts (Runciman, 1974). But many simple concepts are open-textured (Gray, 1977, 340; Miller, 1983, 41–3), and some supposedly contestable ones are not (Gray, 1979, 392–3). This type of non-decisiveness is a form of underdetermination. The second, which is more important for the claim of essential contestability, requires a plurality of normative viewpoints. In this form, the 'logical' nature of the claim for essential contestability is further weakened, for an empirical assumption underlies it; there are many moral theories and there is no empirical way of choosing between them. If people cannot agree over their moral theories, or at least cannot agree on the precise ordering of values within a moral theory, then *necessarily* they will not be able to agree over the correct application of certain words. (This remains a strong claim, for it still involves necessity, but not strict necessity because the prior operator is the contingency of competing moral theories. I do not know how one could try to show the necessity of moral pluralism.) This is because those words are polluted by the moral theories. 'Power' does have descriptive denotation but it also has prescriptive connotations. The way it is used implies and is implied by different value commitments. Connolly (1983, 56) simply says that such words connect descriptive and explanatory statements to normative judgements. It is because they are normative that they are essentially contestable.

I believe that the concept of essential contestability is either trivial or false. It is trivial if it is to be discovered that all words, not just moral or political ones, of any natural language suffer from non-decisiveness or underdetermination. Many, if not all, non-political words suffer from non-decisiveness because of empirical underdetermination of theories. Essential contestability is false if any difference between political concepts and natural ones does not distinguish them as more contestable.

If we want to know what a word means we look it up in a dictionary. The dictionary will give us the conventional meaning or meanings of that word in our language. Living languages change. Dictionaries need to be updated. We discover new facets of the world and the scope of words increases to cover those new facets, or new words are created where the old ones will not suffice. As our science becomes more precise the opposite process may occur and the scope of a word decrease. Here its meaning becomes sharpened in order to capture a small part or mere complexion of a complex reality. Or words may fall into disuse because it is felt that their use hides rather than reveals, or just because conventions change and new words take their place. But when we go about the business of conceptual analysis we are not looking up words in a dictionary, nor even writing one. We are trying to clarify an aspect of our theory on a particular part of that complex. The dictionary writing comes after the conceptual analysis. The lexicographer's lexicon of truth is that his definitions of a word truly capture it as it is used, not that this use itself makes sense given the rest of the users' language. The conceptual analyst's lexicon of truth is that her definition of a word fits into the theory under which she operates; and that the theory she uses is consistent, coherent and, if this adds more, is itself true.

It may be argued that definitions cannot be defended decisively where the test of coherence and consistency for theories is itself indecisive. That is, if the empirical data underdetermine the theory, then concepts in the theory can never be decisively defined. If the data underdetermine the theory then we may not be able decisively to test some of the implications of the theory, which in turn will not allow us decisively to demonstrate preferred formulations for some of the concepts. This is not just a problem for moral or political terms. It is an epistemological problem for many concepts in many theories about the world. Often the evidence that would show one side to be correct and the other false is not available to us and never will be, and this is just as much the case for some propositions in the natural sciences as it is in the moral sciences. The cause of certain evolutionary changes, for example the precise causes of the extinction of the New Zealand moa, may never be settled conclusively because the evidence that would have been available at the time is lost; certain propositions about the nature of the universe beyond our perceptual range may not ever be decidable. Popperians console themselves with the proposition that the debate is scientific and empirical if the hypotheses are verifiable or falsifiable in principle (Popper, 1972, 42). This only produces a problem for those who hold a verification theory of meaning (see below). The lack of decisive argument for one concept over another on the grounds of lack of empirical evidence is not therefore grounds for a case for essential contestability of political concepts.

Anti-positivism is the major ground for the claim of essential contestability. Positivists hold that there is no reason why a dispassionate study of society should not result in a common theory which can be used as a battleground for competing moral and political philosophies. Anti-positivists suggest that this is nonsense since any theory about society is already value-laden. The battle commences before a common theory can get started. However this anti-positivist stance can be accepted without our believing that there are terms over which there *can* never be common agreement. We would only be forced to conclude that there could never be common agreement if we were unable to understand the others' theories, and thereby misunderstand their use of the 'common' words. If, however, we can understand others' theories, and can understand their use of the 'common' words, even though they differ from ours, then we do not have words that are essentially contestable. For all we *need* to do is to mark each 'common' word that does a different job within each theory with the theory it applies to. Thus if we have three theories which disagree over the correct application of the word 'interest' then we can mark that word 'interest$_1$', 'interest$_2$' and 'interest$_3$' in much the same way as Lukes points out the competing views of interest (see Chapter 3). When a theorist of the first school states that we should do x (a word over which there is common agreement in all three theories) because x is in i's interest, a theorist of the second school can translate that as 'we should do x because x is in i's interest$_1$'. She can then reply that doing x may be in i's interest$_1$ but that hardly recommends it, especially since it is not in his interest$_2$. He should rather do y, which is in his interest$_2$. The theorist of the third school will likewise disagree. The debate may be tedious but there is no question of essential contestability over the three words 'interest$_1$', 'interest$_2$' and 'interest$_3$'. Those who argue for essential contestability on the grounds that all concepts are theory-laden forget that all theories are language-laden. The same word may indeed name different concepts within different theories, but those theories are still expressed in a language. The final question is whether or not that language is ever untranslatable into any competing language.

It could still be objected that there remains essential contestability over the term 'interest' unmarked. Each theorist disagrees over its correct application, and its correct application depends upon the truth or falsity of the three theories. As we are unable to discover the truth or falsity of the three theories (if we are relativists we do not think they have truth values) then we will never agree over which understanding of 'interest' is the correct one. But if we are relativists then we cannot ask the question: 'Which of the three ways of understanding "interest" is the correct one?', since we do not think that moral theories have a truth value. Questions of correctness do not arise – though some moral and political theories may still be falsified: ones which rely upon false empirical assumptions, for example. If we cannot ask which of the three is the correct interpretation of the 'common' word 'interest' then we cannot say that there is one common concept, just three words which we all agree have different meanings. There is no essential contestability.

If one is an objectivist or moral realist then one holds that moral theories do have a truth value. Of the three moral theories only one, at most, can be true. The others must be false. Thus the truth of one of the theories gives the correct meaning of the word 'interest'; the other two therefore ultimately drop out as nonsensical, or at least as having no correct application to the actual world. But we could hold that, whilst moral theories do have a truth value, we have no way of finding out what that truth value is. Thus we can ask 'Which of the three ways of understanding "interest" is the correct one?', but we can never answer the question because the data underdetermine. There is no simple answer, for one cannot just go out and look. However, in order to distinguish the theories, we must be able to point at some empirical difference between them. Only if we can understand that theories are different can we perceive that there is a problem. But we can only see that the theories are different if we know what evidence *would* show that one was true and the others false. There are limits to metaphysical speculation after all; even if the positivists were wrong to put those limits on the actually verifiable. But if we can understand what would show the truth of one theory and the falsity of the other, then we do not have essential contestability of concepts. We have contestability in the actual world, but we do not have contestability in all possible worlds; for in some of them we can see which theory is true and which false. For *under realism* the claim of essential contestability can only be one of necessary contestability and thus contestability in all possible worlds.

The thesis of essential contestability does not require relativism. A realist position which accepts value pluralism may sustain a form of essential contestability (Grafstein, 1988). Competing moral systems are possible under forms of realism. This is moral relativity as opposed to moral relativism. Moral relativity is where we understand that the same proposition may take a different truth value under culturally specific circumstances. A particular form of killing may be morally just under one morality but wrong under another. There is no contradiction here as long as we understand that the contrary propositions are contextualized to the speaker's viewpoint. It is not more contradictory than a man on one world looking at a woman on another world and suggesting that his world is stationary whilst hers is moving away from him. She, on the other hand, believes that her world is stationary and his is moving away from hers. A person on a third world may believe that they are both moving away from each other and from the stationary third world. Movement is relative, and it may be measured from any point which may be called fixed, but there are no grounds for arguing that in space there is any point which has a better claim to being fixed than any other. A realist may well accept that there is no fixed point by which to judge all moralities, but does not accept that they cannot therefore be judged at all.

We may distinguish relativity from relativism: the relativist wishes to conclude from the fact that the same proposition is apparently both true and false that we cannot apply truth tables to such propositions; the realist argues that we can if we contextualize them. Admitting the possibility of truth values to such propositions does allow us to criticize the use of moral claims in any moral theory. We do this by trying to show that the theory makes inconsistent claims. Inconsistency does not make sense without truth values.

In actual fact, given theories with unknown and unknowable truth values, we are no worse off than with relativism. We can still talk about 'interest$_1$', 'interest$_2$' and 'interest$_3$' ending confusion about what 'interest' really means and any contestability of concepts. We can agree to disagree over the use of a word, and can continue our real argument over theories about the social world and our moral commitment.

Of course the manner in which I have approached political power brings with it a set of moral concerns. It is not morally neutral in any sense. But this non-neutrality does not entail any lack of objectivity. The arguments may be criticized on logical or empirical grounds and some of the assumptions may be challenged. However I do believe that the rational choice approach brings the conflict between competing theories about the distribution of power into sharper focus, thus allowing the possibility of resolution of some of that conflict. It demonstrates the questions which may be answered by empirical analysis, and those which lie beyond a reasonable expectation of such empirical verification.

8.3 Rational choice and political power

The fact that all political concepts are theoretically and morally laden means that political argument is never fought on neutral ground. We either play at home and defend our theories against others, or we play away and attack others' theories on the grounds of incoherence, inconsistency or empirical falsity. In this book I have fought both home and away, but mostly at home. I have been mainly concerned to defend a particular method of social scientific research and its approach to political power. Whilst this is not morally neutral, it does allow us to bring a new perspective to old debates. I dismissed the concept of structural power in Chapter 1, but rational choice does not discount structuralism or approaches to power which place the burden of explanation upon structures in society. Through rational choice modelling we can see how the relations between actors help to determine or structurally suggest courses of action to actors. Strict structural determinism is ruled out by my assumptions, but rational choice does allow a great deal of situational determinism. The assumption of free will forces us to study human behaviour. One of the great divides in the power debate has been between the behaviouralists and their enemies. I hope I have shown that much of what was thought to be at issue in that debate is unfounded.

Behaviouralism as it has been understood and practised is flawed, for behaviouralism requires a coherent theory of action which was missing. This failure to understand the nature of human action does not require the abandonment of behaviouralism, nor a retreat to non-empirical assertion or metaphysical speculation. The critics of behaviouralism have been correct in pointing out that we cannot assume that actors have a potential power which they could use just because they do not attempt to act. We cannot assume that people do not act for the sole reason that they have no desire so to act. But that is not unbehavioural once we understand and take into account the collective action problem. Explaining behaviour requires making assumptions about interests. I have demonstrated that some of the conclusions of the critics of behaviouralism are quite compatible with behaviouralist assumptions. Both sides failed to take sufficient account of the collective action problems. Rational choice is able to specify the limits of individual action by modelling the structure of individual decision-making. It may also reveal how individuals can transcend structural barriers, even if such transcendence requires heroism, irrationality and luck.

In order to find a method of mapping the power structure in society we need to make the analytic distinction between outcome power and social power. By first studying the outcome powers of actors we ensure that we take into account the difficulties of collective action and do not, too soon, blame our lack of power upon the power of others. Groups of individuals who share common interests must bargain strategically with each other in order to promote those interests. Once we have understood the difficulties of action outside of contrary forces we can then build these in. In this way we avoid the blame fallacy without ignoring social power, which I take, in the last analysis, to be the most important element of political power. This is a strategy for studying power, however, and the distinction is not meant to divide the real world into natural categories. The outcome powers we have are affected by the social powers of others. Real collective action problems involve not only the strategic problems of the outcome powers of groups of individuals working together but also barriers which contrary interest groups erect in their path.

One of the major fallacies pervading power debates is the equating of who has power with who benefits. Brian Barry's concept of luck was introduced to illustrate that fallacy. We often think that the lucky are also powerful (which they may be, but their luck does not demonstrate this), and this is explained by systematic luck. Some groups in society always seem to get what they want, at least in some policy areas. To equate this with their power is an understandable but serious error. It is a serious error, because to try and discover the power of the systematically lucky is to search for something which can never be found. It is this search by the non-behaviouralists which renders their work non-empirical and non-behavioural. Because they could not demonstrate empirically what, logically, they cannot discover empirically they came to scorn empirical research itself. But it is possible to uncover systematic luck. We can discover it not by searching for the resources and the actions of the powerful, but by looking at the structures in society which lead different groups, state actors and governments to formulate the policies they do. This is structuralism at its strongest, but it is not non-behavioural: not once we have understood the nature of human behaviour.

This book is a conceptual exercise. I have tried to demonstrate fallacies in previous accounts of power and to produce a coherent and consistent account of my own. It is empirical in the sense that I have illustrated my account with examples of empirical research to try to show how useful the conception offered here is to empirical political scientists. I have not tried to conceal my own views of the existing structures of power in Britain and the United States; nor have I tried to give a systematic account of those structures. I take it that pluralists, elitists and statists can all use the rational choice method. Analytic marxists have already demonstrated its usefulness for marxism. Using rational choice methods will affect the forms these different theories take, but the method itself will not generate the thesis which will prove the most empirically robust.

PART II
Postscript[1]

9

Some Further Thoughts
on Power

9.1 What I would do differently now

Following its publication, the reaction to *Rational Choice and Political Power* (hereafter *RCPP*) was, to put it mildly, rather slow. Jim Johnson, whom I did not know at the time (I later co-edited the *Journal of Theoretical Politics* with him), wrote a nice short review in *Ethics*, Peter Morriss a longer and more critical but respectful one in *Utilitas*, and there were a handful of other reviews – in *Political Studies*, *Acta Politica* and such places – but that was about it.[1]

I realize now that I blew it with the title. Most academics who were interested in social and political power looked askance at rational choice. Meanwhile, rational choice scholars who thought about power were interested in work far more formal than that contained in the book. What I wanted to do was to bring the results of formal work to bear on the empirical analysis of power in society, whilst critiquing those who did not think we could study power empirically. By keeping it non-technical I thought I could draw the power researchers in, but my title put off most of those who might have cared. The original idea for *RCPP* came from my critique of Steven Lukes's three dimensions of power. I thought the collective action problem could explain everything which he thought required adding dimensions of power. However, *RCPP* does not simply analyse power in terms of collective action but uses, informally, a decision- and game-theoretic framework to conduct its conceptual exercise. I argue that we can *measure* power by agential resources, building up a picture of the power structure through resource-holding and the incentives that lead some to be privileged beyond what their simple resources seem to

173

suggest. And if I had used the term 'privileged' for those who gain and 'wretched' or similar for those who lose, my argument there might have drawn less criticism.

Having said that, if I was rewriting the book today, I would make collective action even more central, since it can help explain elements of the power structure outside even of Lukes's general account. I also believe that game-theoretic analysis of the internal mind – how we think of our own identities – can illuminate aspects of social power that go beyond conscious interpersonal relationships. A few years later, I published another little book: *Power* (Dowding, 1996a). It contained some of the central arguments of *RCPP*, but deliberately made no reference to the work of Steven Lukes (1974, 2005); it suggested how preferences can be formed through inter-relationships. That book attracted the attention of many undergraduates for a while, but again had little effect on the literature. I still believe that thinking about one's own identity and beliefs through how society views one, given how society structurally suggests identities and preferences, can make the collective action problem even more central to our analysis of power. So the first thing I would do differently now is to call the book *Collective Action and Political Power*.

I would change a few other small things. I would not now describe 'power to' and 'power over' as 'outcome power' and 'social power' – largely because I think the overall topic is best couched in terms of political and social power. Indeed, in the two chapters of this Postscript I will use 'social power' in this broader sense, and not as how I defined that term in *RCPP*. The 'power to' do things along with others – collective action or what is sometimes called 'power with' – is as social as any other form of power. Now I would stick with the terms 'power to' and 'power over', seeing both 'power over' and 'power with' as subsets of 'power to'. Alongside these small changes, I would take some of the discussion a little further. This is the task of the postscript.

9.2 Rational choice and collective action

Rational choice theory has long been disparaged by the majority of social scientists. In part this is ideological – economic methods were the preferred choice of the right, especially those designated 'the new right' in the 1970s and the 'economic rationalists' or 'neoliberals' of more recent times. Rational choice does indeed engender cynicism, with the knowledge of how political processes, including democracy, can become corrupted. However, what you get out of a method, generally speaking, depends on what you put in.

Critics continually attack rational choice for its assumptions of self-interest and instrumental rationality as being the most important form of rationality. In fact, preferences and von Neumann–Morgenstern utility functions have no content – content is provided either by assumption or by the interpretation of agents' behaviour. In other words, for empirical analysis, what goes into utility functions is empirically determined. Theoretically we can put in anything we like. The detractors miss the point that agents can maximize over whatever they like.

Token biological humans are rarely studied using rational choice methods (I suppose 'analytic narrative' approaches are a partial exception: Bates et al., 1998, 2000). Instead we study *types* of individuals. So, from the behaviour of types of people, we learn what enters into the utility function of those types when they are acting in their roles with regard to their type. Of course, biological people belong to many types, and hence trade across what we theoretically maximize within their types. The content we put into utility function by studying types of people in one situation are then used to predict their behaviour in new situations. Those scientific predictions form the focus of the explanation (Dowding and Miller, Forthcoming), usually driven by the incentives we see in diverse situations – that is, the structure (Stigler and Becker, 1977). (I discuss the nature of this exercise in more detail in Chapter 10.) When doing normative theory, we can put any assumptions we like into a utility function, and sometimes surprising results emerge. For example, we find that altruists and the virtuous face collective action problems under certain conditions, as long as they want their actions to be used most effectively for whatever it is they judge to be the right thing (Schofield, 1985).

The three rationality assumptions of revealed preference theory provide consistency for prediction and have absolutely nothing to do with rationality beyond that. If you have weird beliefs, but abide by the rationality assumptions, then we can predict the most wildly irrational behaviour that we can imagine any agent conducting. Under certain strategic conditions, random behaviour is rational. These 'rationality conditions' are required, in my view, in order to interpret what people are doing – so any theorist who makes any assumptions about explaining human behaviour in terms of reasons is a rational choice theorist whether they know or like it. In reality, though, the 'rational choice programme' of *RCPP* is made up of two simple claims. The first is that Lukes's need for three dimensions of power evaporates when we understand the collective action problem. The second is that if we can measure agents' resources, then we can measure their power. To the extent that we can measure the resources of agents we can give a comparative statics measure of agents' power. In doing so we also map the power structure. If we want to give a dynamic analysis of actual (token) political process, then we have to work out preferences, we need to take account of reputation and expectations. In Section 9.3 I will talk about resources. In this section I will defend the collective action claim.

RCPP (ch. 5) argues that in the classic community power studies, contrary to many claims, we can see that both capitalist agents and the people within a community act to promote their divergent interests. Since capitalists have more resources and their interests are concentrated, they do not face the collective action problem that citizens, workers, consumers or voters do. Offe and Wiesenthal (1985) described this as the 'two logics of collective action'. In other words, we can explain the amount of activity based on the incentive structures of all agents. The major point I try to make is that we do not have to show that another agent is working against oneself in order to explain one's lack of power. Examining why people do not work together ('power with') can explain a large part of powerlessness. We do not *need* to invoke the power of others.

Now, of course, we can always ask the counterfactual question. What if the large group *G* were to collectively act against the interests of the small or privileged group *P*? Often, if not usually, *P* would respond. Indeed, one of the strategic considerations of the members of *G* is the likely response of *P* and the subsequent consequences. The fact of *P* is part of *G*'s collective action problem. In other words, claiming that *G* faces a collective action problem, and that is why its members do not act in ways that might seem to be in their interests, is not to deny the power of *P* to stop them were they to try. Nor is it to deny the weak position of members of *G* in relation to members of *P* or to the relative powers of the groups once mobilized. The collective action problem is invoked simply to show why lack of activity on the part of members of *G* does not, *in itself*, tell us anything much about their interests or their relative power in relation to others. At the very least, we have to strategically model their situation before we can conclude anything about their interests and their relative power. The collective action problem can help us to explain the acquiescence of the weak within the structure of society. It can help us to model the power structure. And, I maintain, in order to strategically model it, the best place to start is to look at the relative resources of all the agents.

We ought to be able to understand the importance of the collective action problem better now than when *RCPP* was first published. At that time, one of the issues that most vexed those doing power studies is why some agents act in ways that do not seem to be clearly in their own interests. That question still vexes. But subsequent work on the collective action problem and toy games like Prisoners' Dilemma have illuminated it. Even repeated coordination games can lead to failure, especially where the game has suboptimal Nash equilibriums (Parkhurst et al. 2004). Alan Carling (1991, ch. 7) uses Chicken games to explain how two types of people (say men and women) can be locked in to situations where one type exploits the other. Gerry Mackie (1996) uses toy game theory to explain how suboptimal conventions can arise in issues such as foot-binding and infibulation. More generally, complex agent-based models show how suboptimal outcomes can emerge for groups in iterated game situations (Bruner, 2015; LaCroix and O'Connor, 2018; O'Connor, 2019). Many of these arguments are not directly about power as such, but they are designed to show how oppressed groups can acquiesce in their own oppression. Others explicitly working on power and oppression have used rational choice theory to explain how those who are oppressed can acquiesce in their own oppression. John Heath (2000) suggests that the excessive use of beauty products and efforts to achieve a prescribed feminine body shape can be modelled as a collective action problem where each individual woman would be better off not engaging in the oppressive practice, but worse off in social life unless all abstain.[2]

Anne Cudd (2006) develops a rational choice argument to explain how collective action problems lock in suboptimal choice for oppressed groups. She argues that we can see agents choosing to maximize their utility within socially structured payoffs that are the result of previous, equally rational choices. So the choices of the oppressed are individually rational – they are the best short-term choice in those individual circumstances – but are not in the long-term interests of the group as a whole. This argument is applied to the gender division of labour in the household (see also, for example, Manser and Brown, 1980; Becker, 1981; McElroy and Horney, 1981).

Now we must note that none of these arguments claims that the collective action problems of oppressed groups cannot be overcome. Rather their critical, radical message is that if we are to overcome oppression, these collective action problems need to be addressed. Those facing them need to recognize the problem and strategize how to solve it. Or we can look to altering some of the structures of society in a top-down manner in order to enable the oppressed to fight against this oppression and their rational acquiescence. We should also understand the high-level generality of my and others' arguments. We use *the* collective action problem to illustrate the issue, but in fact every group faces unique collective action *problems*. The Prisoners' Dilemma, Chicken and the like are 'toy games', useful for illustrative purposes only. One has to utilize real game theory to fully model actual problems. We must not forget that

Lin Ostrom's Nobel Prize was for her work on common-pool problems (a type of collective action problem); she showed that history, convention and expectations provide different solutions for diverse common-pool problems (Ostrom, 1990, 2001, 2005).

Can the collective action problem really model all aspects of oppression, however? Of course not, and I do not argue so in *RCPP*. Collective action can show us why acquiescence can be rational (not that it always *is*). And we must not think that because the weak acquiesce that is the full reason they are oppressed – though, generally speaking, the collective action problem is the most important element. If the weak resist, the powerful will flex their resources to keep them under – if that is what the powerful want.

Amy Allen (2008) suggests that there are limits to both rational choice and collective action accounts of oppression. On her general critique of rational choice theory, I will say two things only, then move on swiftly. First, there are many legitimate methods of social science study, and they are not necessarily rival. Rather, they can provide complementary explanations of different aspects of the same phenomenon. Indeed, some might be directed at type-level explanation, others at token-level. Rational choice analysis, like discourse analysis, is a method; methods are not falsifiable, simply more or less useful (Dowding, 2016a). One cannot prove the uselessness of rational choice by showing that a particular rational choice model fails to explain every aspect of some social phenomenon. Nor can one discount discourse analysis or any other approach in this way. Of course not, because there are rival rational choice models or rival discourse accounts of the same phenomenon. There are different models, or arguments, within each method that might be rival, and there are good and bad accounts of social phenomena within each method. Second, Allens' general critique of rational choice falls into the trap mentioned above of misunderstanding utility functions and the importance of the type–token distinction. Nevertheless, she poses an important challenge which I want to address.

Allen's (2008) critique of Cudd is that her argument does not fully explain the nature of oppression. Cudd argues that four conditions characterize oppressed groups. First, oppression occurs because of membership of a given group. Those members are subject to unjust burdens, discrimination or constraints. These harms result from coercive practices. Finally, other groups receive benefits as a result of the harm meted out to the oppressed group. However, what is most problematic to explain is the enduring quality of oppression and how the group members respond to it, often acquiescing in their mistreatment. What is original in Cudd's account is not simply that it is often rational for agents to acquiesce, but the insight that their interests in so doing are structurally created by the situation in which they find themselves.

Allen illustrates this structural interest with a story about a married couple (Luke and Samantha) who find it in their joint interests to see his career develop and hers not, because his earning power is greater. So Samantha becomes economically dependent on Luke. The structurally induced interests come about because the woman has chosen a career – in Allen's example in primary education – which pays less than her husband's career in engineering. Women tend to choose such careers because they better enable them to combine a career with traditional female domestic duties (Okin, 1989). In other words, historically induced interests cause women to pursue a path that leads to continued oppression of their type in the future. The collective action problem of women as a type is that, in order to change paths, they need as a type to decide not to act in this manner. Women would have to break out of the current statistical distribution of job types and domestic duties, to achieve a distribution equal to men. *Even if* there was no pushback from men, as a type, to that shift, such a collective action is problematic. First, there is no obvious mechanism to bring about a fast shift in career choices for women. Second, their preferences for career type are structurally suggested, by the norms, conventions, expectations and education they imbibe at home, at school and socially.

Now we note here that we require two different models to examine Allen's narrative. The first models the choice situation of Samantha: that is, the decision of a token person given the incentives she faces. In that model we can take her preferences and her choice situation as given. We can then see how it is she makes a decision that seems reasonable under the circumstances, even though it means she falls prey to the oppressive problems of women in her society. Next, we can ask: where do these preferences come from? To answer that, we utilize a second model – one that explains how it is that women, and men, in general take on the attitudes they do in their society. This is a type-level explanation. Such a model is going to be much more complex, but the simple version of it suggests that certain attitudes are formed due to expectations, norms and conventions. These expectations, norms and conventions can be modelled as long-run equilibriums of repeated games. To break away from them requires either the solving of a collective action problem – a host of people must agree to act in ways that lead to a new set of expectations and so create new norms and conventions – or some people playing out-of-equilibrium strategies that disadvantage them, but thereby inspire others to change their behaviour and so create new norms and conventions. The latter sorts of models use evolutionary game theory, where stable equilibrium comes about following equilibration, where the players are types and the payoffs are to types rather than individuals. The point is that we have two separate models, one using token players to show how it is individually rational for a person to behave as they do and a second with type players explaining how attitudes are formed.

Allen critiques the argument that as women increase their investment in careers and engage less in domestic duties, they should begin to earn more, whilst we should see men taking on more domestic duties, especially in households where women earn more than men. She notes that various writers recommend this as an individual strategy (Mahony, 1995; Heath, 2000). However, empirical evidence is more complex. As women's earnings increase up to parity with their husbands, division of labour in the household does become more equitable (Bittman et al., 2003; see also Tichenor, 2005). Nevertheless, where women earn more than men, the opposite occurs, and men do less household work. To the extent that higher-paid women do less household labour, either it is left undone or they purchase domestic labour.

What does this show? Allen and others are certainly correct that gendered norms are coming into play. Following Bittman, Allen suggests that women earning more is a 'gender deviance' and the household compensates for that by a more traditional way of dividing housework. We have to be careful. None of the empirical evidence suggests that in any actual token household, housework is divided more evenly as the income of the two partners becomes more equal, and then as the woman earns more the man does less. Rather, households closer to income parity are closer to parity in domestic work; and in households where the woman earns more, men do even less of their share. We need to consider what the *types* are in these households. I would suggest that in households where women earn more – because they are professionals and their husband is blue collar or, in working-class families, because the woman has a regular job and the man does not – the traditional expectations are more likely to be held by the men. What we are seeing is social expectations based on class divide.[3] Furthermore, I would expect that men in that situation would use those expectations as resources in the implicit household bargain, as indeed Tichenor (2005) reports.

I do not disagree with Allen, Bittman et al. or Tichenor in their analysis of these situations. Norms are driving the expectations. However, I do not think that these norms are in any way contrary to the model representation of the domestic power game within the token household. And furthermore, I think that we can model the type power game as a repeated generational game that can explain how the norms operate in different types of household, and how we can expect to see those norms alter over the generations. Again, we must be careful not to confuse type-level explanation with token-level, nor to confuse rational choice models at either level, nor to think that other types of explanations are necessarily in conflict with the rational choice ones.

The collective action problem does not demonstrate every aspect of the power and privilege structure. However, it forms a vital part of explanations for why some people (types) acquiesce in certain situations and why token individuals do not always act in their apparently best interests. We also need to look at the resources that people bring to both explicit and implicit bargains, which include the norms and expectations that others bring. I agree that norms and expectations about masculinity and femininity play a large role in the oppression of women. However, I think it is incumbent upon us to explain how these norms and expectations arose, and how they are used both intentionally as a power resource and unconsciously in ways that instantiate privilege.

9.3 Resources

RCPP provides a resource-based account of power. It does not argue that power is a resource that is used by agents (Okin, 1989); rather, it argues that we can measure agents' powers by the resources they command. And we can map the power and privilege structure of any given society through that measurement. I write 'power and privilege' structure here, because of my argument that some groups get what they want without trying. They are systematically lucky. Consequently, other groups are systematically unlucky. The systematically lucky might be thought to be privileged (though we should not conclude that all those who are systematically lucky in some context would, overall, be considered privileged). Most writers on power talk about the 'power structure' and would include within that both those who gain through the power they have (by my account) and those who gain through luck (by my account). Strictly speaking, what they call 'the power structure' is by my account 'the power and privilege structure'; and for me 'the power structure' is a subset of the broader structure. The labels are irrelevant, really; what matters is whether my analysis partitions in a manner that is explanatorily and/or normatively important. I think it does.

We cannot simply read off agents' power from the resources they have, because we have to look at the context in which they operate. We have to look at others' resources too. Furthermore, agents might not use the resources they have; even if they are powerful in some contexts, they might not appear so. Given that agents often have to work together, they might not be able to use their resources effectively, despite any individual's best efforts. Nevertheless, we can analyse and measure agential power by their relative resources. My account uses the game-theoretic work of John Harsanyi, describing four ways in which power operates. They are all outlined at a high level of abstraction.[4] If I were writing the book now, I would go into more detail about how we could actually go about measuring power empirically – that is, the resources we would measure within Harsanyi's four ways and a fifth (reputation) that I add.[5]

Whilst the resource-based account is agent-focused in one sense, it also provides a structural account of power. The power structure of society is determined by those relative resources and what counts as a resource is often socially determined. In Chapter 1, I argue that agent-based 'methodological individualism' and structural or holistic accounts are not in conflict. Indeed, I thought the methodological individualist–structural argument was over and everyone could see that one requires the other, so I did not say much about it. So naïve! I will say more about this in 10.2 and 10.3.

Harsanyi's account of power (1962a, 1962b, 1976) that underlies my conception was an explicit attempt to formalize Robert Dahl's (1957) original definition: 'A has power over B to the extent that A can get B to do something that B would not otherwise do'. This is clearly a version of 'power over'. However, I see power over as a subset of 'power to'. This is very important, in my view. I think the entire debate over whether social power should be seen as 'power to' or 'power over' is a complete waste of intellectual effort. We have a general term 'power', and we can subdivide it into the power that agents have on their own, together with others, and how they can affect others. They can affect others directly and indirectly, and this is important both normatively – in terms of blaming people for what they do – and critically – for how we want to change society. Logically, it seems to me, 'power with' (Follett, 1942; Allen, 1998, 1999; Pansardi, 2011) and 'power over' are subsets of 'power to', because the general understanding of power is getting stuff done – which is 'power to'. Analytically, that it is.

Whether 'power with' or 'power over' is more important depends on the questions one is asking. The community power literature was certainly about 'who governs'; that seems to be a 'power over' literature. The empowerment literature is largely about 'power with', and thus a form of 'power to' according to *RCPP*. Domination is about 'power over', though when it touches on acquiescence it is about 'power to' again. So I see little point in abstract argument over the importance of what type of power matters outside of their specific context.

One problem with my account is that, whilst I am clear that conscious intention is an important component of how we normatively assess actions, conscious intention is not important in assessing an agent's power. It is normatively important whether someone's acts are consciously racist or sexist, or whether they deliberately hurt others, but what they do is still an act of power, conscious or not. Moreover, I did not say explicitly enough that luck and systematic luck are part of the power structure, or what I would prefer to call the power and privilege structure. Being privileged or wretched may not be as a direct result of any agent's action as such, but being in such a position is directly attributable to the structure of society. In my view, in order to promote social justice and emancipation these distinctions are vital.

Information

Using information is obviously one way in which power operates. Holding some information and having the means to utilize it is a power resource. Principal–agent models of economic and political resources rely on asymmetries of information. The principal–agent relationship is based on agents having information about what sorts of people they are and how well they are carrying out the tasks the principal sets. Representative government can be seen as a chain of delegation from voters, through parliament or legislatures, to the elected executive and on to the non-elected public service executive. Of course, we also have such relationships within firms, and between people and those with whom they interact, such as retail outlets, restaurants, plumbers, doctors and so on. In these relationships the agent holds informational power over the principal; the principal has other powers to help overcome problems that might result from informational asymmetries. A well-functioning market does not stop an agent from cheating or reneging on each individual deal, but it does give incentives for agents to be honest, lest their dishonesty in relation to other traders causes loss of business. Simple textbook accounts of the welfare efficiency of markets usually ignore information costs, but we are aware that information seeps out through myriad market transactions. Nevertheless, we have laws, albeit imperfect ones, that attempt to control the worst abuses of informational asymmetry in markets.

The principal–agent model is a model. Fully specified, it gives us some predictions, and it suggests ways in which principals can try to overcome the agency problem they face. Parliamentary and legislative procedures provide scrutiny of politicians and state actors and, with electoral incentives, give reasons for politicians to be careful. The legal system, ideally independent of the political system, provides further oversight and its sanctions also provide reasons for agents to act in ways acceptable to the community. And, of course, the media has oversight over all the elements and conveys information to the public. All this helps to overcome the asymmetries of political information between the public and the political masters. It does not stop powerful interests from having greater input into the political system than ordinary citizens, but is supposed to provide some check upon it.

We have seen recently that this system relies far more than most people imagined on the moral qualities of agents and their acceptance of the norms of rightful behaviour. Donald Trump and the Republican Party in general have flouted previous norms and simple morality to an extent unimagined by even the most seasoned experts in US politics. Trump has demonstrated how cynical and flagrantly unprincipled behaviour can be maintained if one does not care how history will judge one and one promises material rewards to a large coalition of support. He has shown that blatant lying will be accepted by supporters who want to believe their heroes and are supported by venal news outlets and the ubiquitous force of social media – we are only now learning how ubiquitous and effective the force of social media bots truly is.

These things are measurable. We can study the causal effects of differential media coverage, we can track down falsehoods and inconsistencies in claims. We estimate readership of different forms of media, and can see the average effects upon readers or listeners. We have measured how framing affects the way in which people view issues, and also how people choose frames (for example, Chong and Druckman, 2010; Druckman et al., 2012). The latter suggests that information holders are not simply powerful people, but they can also gain traction through luck: that is, we can estimate the numbers of people whose preferences already map on to the views of those who control information.

One challenge to my view of power is that one resource that we have is our ability to persuade people. I can give or withhold information, and I can give it in different forms. We tend not to think of a debate between two people as being a power game, but of course it can be. In any discussion one person can choose to withhold information from another, to lie, or present information in ways that might persuade. By Dahl's (1957) underlying account of power, these tactics are descriptive of a power relationship. I try (Dowding, 2016b) to give an account of when such acts of persuasion are not coercively powerful or manipulative. This comes down to two semantic conditions:

> *Common Reason*: Roughly speaking, if an agent i induces individual j to assent to S on the grounds E, and j's assent to S is on the same grounds E as i's assent to S, then i has not

manipulated *j* into assenting to S. The grounds E constitute a common reason for assenting to S for both *i* and *j*, even though the process through which *j* assents to S is *i*'s establishing E for *j*.

The idea is that if both people come to believe S because of content E, then E constitutes content-dependent reasons for both of them, rather than the persuasive ability of *i*.

The second condition is:

> *Deliberative Honesty*: This is about the intentions of *i* when she makes the statements that induce agent *j*'s assent to S. If *i* persuades *j* of S on grounds E with the intention that together they learn the truth or, in other contexts, come to some agreement, then *i* has not manipulated *j*. If *i* intends to persuade *j* of S no matter what and is prepared to use any type of rhetorical strategy to achieve that aim, then *i* is manipulating *j*. Furthermore, *i* is manipulating if she lies or strategizes to induce beliefs, even when she is unaware of agent *j*. Agent *i* still wields manipulative power if, hundreds or thousands of years prior to *j*'s birth, she makes the statements that induce beliefs in *j*. What matters is *i*'s veracity at the time of making the statements.

The idea is that if the two are involved in a cooperative exercise then 'power over' is not operating. Instead, they are using their intellectual and informational resources together, in a form of 'power with', to understand some aspect of the world or come to some agreement.

Dowding (2016b) is all about rational agreement and ignores an important aspect of human discourse and interaction: emotion. In fact, the Humean distinction between reason and desire, or rationality and emotion, which might have some analytic purchase, is ontologically misleading. An emotionless person appears completely irrational (Damasio, 1994; Bechara et al., 2000). So I address the same question with regard to emotions in Dowding (2018). The question here is when is the use of emotions manipulative. I give two conditions mirroring those above:

> *Common Cause*: Roughly speaking, if an agent *i* induces individual *j* to assent to S at least partly through the induction of emotion F, and what induces F in *j* is also what induced emotion F in *i*, then those grounds E constitute a common cause of F in both *i* and *j*. Or the induction of emotion F in *j* is a result of *i*'s semantic actions under semantic condition (2).

Either way *i* has not coerced or manipulated *j* into assenting to S.

The emotions that are generated in *j* might not be the same as in *i*. For example, *i* might be fearful and *j* angry, but the idea is that the cause of both sets of emotions are the same, and *i* explaining her condition is only the process of transference and not the common cause.

The second condition is simple:

Emotional Honesty: The emotions expressed are genuine.

In these cases, though, *i* will have affected *j* and we can say therefore that although she has exercised power, it is not manipulative nor coercive.

These examples show that power is not necessarily normatively negative. Working together can bring positive gains to everyone. Even 'power over', whilst for liberals it might necessarily have some normatively negative connotations, can be justified. An inspiring speech can be manipulative, the rhetorician using all her wiles to enflame and encourage her audience, but that manipulation might be to their advantage and for the overall good of all.

Language gives us a good insight into the differences between those who want to view power outside agency and those who want to see it as something that agents wield. In the account I have given of the use of information as a form of power, the intention of the speaker is important. By intentions I do not just mean conscious intentions. We can intend things without realizing it at the time. That does not make empirically analysing such situations for coercive power or manipulation problematic. It might mean that we cannot do so quantitatively, but when I say empirical examination I do not just mean quantitative analysis. We can investigate interactions and power relationships between agents and judge whether the interactions were intentional or some of the consequences due to unintentional by-products. Beyond that aspect however, some would argue that there is power in discourse without any kind of intentions of the speakers.

This is, of course, the claim of Foucault (for example, Foucault, 2015, pt 2, ch. 1) that language structures our very way of thinking and thus determines our analyses and interests. He also seems to suggest that to use a word to classify a thing is what brings that thing into existence. Taken too literally, this cannot be so, since surely the causal force works in the opposite direction for many objects. We label them because their existence affects us. Nevertheless, if English lacked the word 'rape' or any equivalent, its existence might be harder to identify. We know from Sapir–

Whorf experiments that different names within colour spectra mean that different language communities see different colours (Kay and Kempton, 1984; Roberson et al., 2000). However, experiments also demonstrate that these seem to have only small effects on thought. So Foucauldian claims seem somewhat limited. Worse, if we make our claims about power and domination as broad as Foucault seems to suggest, we have no contrast to make. If domination is so ubiquitous, it loses its normative force (Dowding, 2006).

Bourdieu (1991) likewise suggests that language itself is a form of power, that operates outside of any intentions and for the advantage of some over others. I do not deny these structural effects of language – there is hidden or unrealized bias in the perlocutionary force of language-use – but I dispute the overriding importance that others give to them. The general Foucauldian argument claims too much and, because it does not allow for nuances, will not admit any empirical evidence to disprove it. Handy if you do not want to be shown to be wrong.

Whilst discursive practices might constrain us, language does not completely determine our beliefs and interests. If that were so, then discovery, invention and new ideas about the nature of the empirical world and our moral universe, including about how language might bias our thought, could not emerge. But they do emerge, and not all new discoveries and ways of thinking perpetuate advantage. We are able to break through some of the bias of our language in order to understand truths and seek reasonable agreement. Indeed, through the academy and through ordinary discourse we constantly challenge our natural language. Some people explicitly attempt to change our language in order to express new ways of looking at political and social situations; others do so implicitly and non-consciously. Language is sometimes used for rhetorical purposes; not all rhetoric promotes the advantaged or dominant groups. Indeed, a large part of academic conceptual analysis of the meaning of such terms as 'power' is part of the process to overcome possible promotion of the advantaged or dominant. But I have already fallen into speaking of language as a tool, a resource of speakers, and got away from it being a form of power itself, as do most people who talk about discursive power.

Legitimate authority

There are accounts of power that distinguish it from authority and related concepts. I do not think that there are any (or at least not many) natural divisions in social concepts, and believe that we should make divisions only in order to do an explanatory job (Dowding, 2016b, ch. 8). Distinguishing some form of naked power from authority might be important in some normative or even explanatory contexts, but generally speaking authority can be seen as a power resource.

Authoritative directions can best be seen as giving content-independent reasons for action. This is so whether the directions come from someone who is *in* authority or who is *an* authority. We might do as we are asked by our doctor, lawyer or accountant, even if we do not really understand the reason why. We follow their directions because we assume that they have good reasons for them. So our reasons are simply that they told us so. We do as they ask for content-independent reasons. By Dahl's simple definition, they are exercising power over us. This notion of power is merely descriptive. They need not be exercising manipulatory or coercive power. If they abide by the two semantic conditions, then they have not manipulated us. In my account that is because their reasons become our reasons through our acceptance of their authority and their abiding by the two semantic conditions (Dowding, 2016a, 10–11). Our reasons for obedience are content-dependent at one remove. However, when they do not abide by the two conditions, they are manipulating us: their reasons are not our reasons and we act entirely for content-independent reasons.

Those *in* authority also have great power. Sometimes we should obey directions simply because that person is the agent coordinating our activities. That role gives them authority. Again, their use of that power need not be manipulatory or coercive and we might follow their directives for content-independent reasons. If they abide by the two semantic conditions, then their reasons can become the content of ours at one remove. Having such authority is a major power resource, and examining its character involves establishing to what extent agents have power. We can look at the formal rules to see what authority people have. We can compare the formal powers of chief executives in different countries, for example, and assess their relative authority.

The authority people have in part depends upon the coalitional basis of their support, however. A prime minister who has the full support of a majority party in parliament can utilize their formal powers far more authoritatively than one without. Formal powers can be exercised with impunity or precariously. Furthermore, the authority of many roles in society, both in politics and society more widely, derives from norms, conventions and expectations. Donald Trump has demonstrated that with coalitional support a US president can push the boundaries of executive office much further than many commentators understood. The Republican Party leadership in Congress in Trump's first two years also demonstrated how much of what was understood to be constitutionally required behaviour is based merely on convention and norms that can be ignored or rewritten. We can, though, compare the formal resources held by different agents in a society and by agents in similar positions in different societies. This gives us the *de jure* powers of agents; norms, expectations and conventions give us an idea of their *de facto* powers, which can be extended if veto players allow it.

Personal power resources can also include characteristics or traits of the person. They can play a vital part in how far someone in an authoritative position can push the boundaries of their role. Some people might simply be more persuasive than others. That does not mean that they are manipulating or using coercive power (though they might be; see Dowding, 2016b), but they are still exercising a form of power. Another non-material resource that people have is reputation (see below). They might simply be reputed to be reliable and trustworthy. Social psychology provides many insights into such personal power characteristics. In detailed analyses – of historical token power relationships, for example – individual traits are important and can be a vital source of power. When looking at social forces more generally (which is more the business of political science), personal traits drop out. For questions that involve large numbers of people, we can assume that these psychological traits are evenly distributed across the population in comparative sectors.

Some personal power resources are a result of biases in the way society is structured. We can think of a gendered example (based on Dowding (2016b, 11). Imagine two agents i and k who each believe some mechanical fact S about a car engine. When k imparts information S to j, j does not believe her. When i does so, j believes him, because he thinks i is more likely than k to be an authority on S. Say, the man j assumes another man is more likely to know about car engines than a woman. In this he is simply wrong: k is a mechanic and i is simply repeating something he once heard. We see here that i has a resource (being male) that k lacks. Now we need attach no blame to i because of this – it is not his fault he is male, even if he takes advantage of that fact occasionally. He is not manipulating or coercing others as long as he abides by the four conditions above. His persuasive advantage is, in my terms, luck (and, if society is structured in certain sorts of ways, systematic luck). This luck gives him descriptive power. (I examine the relationship between descriptive power and luck below.)

The example tells us something about the structure of power and domination in a society where this happens. It tells us that the structure of beliefs I've imputed becomes, in this example, a power resource of men (Bourdieu, 2001). It can be extended to other sorts of beliefs, based on prejudices relating to ethnicity, race, education, class or even clothes. My point is that we can map such power inequalities through such perceptual biases, and these biases form part of the power structure. When people intentionally use these biases to their advantage, they are exercising power; when they are simply advantaged by them, they are lucky or, we might say, privileged. Systematic luck is very much part of the structure of society and to the extent that we are concerned about equality we are concerned about the distribution of luck. Of course, that is the argument of luck egalitarians, who want to reduce inequality due to 'brute luck', whilst allowing for inequalities caused by option luck. I argue (Dowding, 2010) that equalization will not only reduce brute luck, but will also reduce the effects of option luck, and briefly consider the difference between luck egalitarians' conception of luck and that in *RCPP*.

Conditional and unconditional incentives to change others' incentive structures

I will say far less about the conditional and unconditional incentives to change other incentive structures than I have about authority and information. Not because they are less important, but because I have less to add to *RCPP*. Conditional incentives are those threats and offers that we can see being made in power games, in bargains, all the time. In order to make those conditional incentives, one needs other sorts of resources – money, access to information, means of violence, whatever. Some of those resources are, of course, those that underlie authority and information, but they can be *used* in a different manner.

When an agent makes an offer or a threat or a combination of the two (a throffer), we can see an obvious power-exchange relationship. We can measure the relative powers of people in this regard by the sorts of resources they have access to: money, withdrawal of support, job offers, offers to work with others and so on. We do not always think of these offers as power plays, but they are a form of descriptive power. In *RCPP* I say that this form of power is overt, as both parties need to be aware of the conditional incentives. These perhaps are the resources which are easiest for us to measure, and the most obvious ways in which we can conduct comparative statics. Money does not always translate into getting what you want. But the reason for that is that success = power + luck. In the resource account of power, what we measure in comparative static analysis of the power of different agents are the resources they could bring to bear in any cooperative or conflictual power situation. Thus a direct measure of agents' abilities to conditionally changes others' incentives to act in one way or another is the relative resources they have available. The main sets of resources that people have for conditional incentives are those material resources they have, alongside information and authority.

Unconditional incentives are where one sets the conditions for choice. The government has the biggest resource in this by making laws and regulations, but one can also think of how employers might set the conditions for their employees. We can think of the language considerations of Foucauldian analysis in these terms too. So both authority and information can enter into models where we are thinking about the conditional and unconditional incentives. The general idea is that we can think about the power and privilege structure in these terms as a way of structuring how we measure power in terms of material resources and the strategies that actors adopt.

Reputation

Reputation is a necessary addition in strategic situations. When we map the power structure by looking at the resources that agents have we will be mapping their capacities. However, there is one capacity that will not be mapped. And that is the reputation of agents based upon their past behaviour. Reputation in my account is meant to fill that lacuna. It becomes of vital importance when we examine real power games in strategic contexts. We can think of reputation as a resource in such dynamic games, and under that heading I include expectations about how each person is expected to behave. This final category is supposed to cover the factor missing when looking at the other general resources. However, as some have pointed out (Barry, 2002), it is a little problematic for my account – though I think not more problematic than for rival accounts. Whilst reputation is needed when we dynamically model actual power processes, it is difficult to utilize in a comparative statics framework. All we can say in that kind of framework is that if our measure of the relative resources of different agents seems to underspecify the power of some of them (even taking luck into account) – they get more than they ought to in bargaining – then that might be a result of expectation inherent in reputation. Reputation is difficult, though I believe not impossible, to measure, even if only roughly. The reputational scores of the first community power studies provide an early exemplar, and more sophisticated interview and survey techniques are available.

9.4 Conclusions

I have said a little more about some of the aspects of my account of power in this chapter. I have defended the idea that the collective action problem can explain most of why groups of people are often powerless and why dominated people often seem to acquiesce in their own dominated status. This does involve thinking about the particular collective action problems of token individuals and the collective action problem of their type in separate accounts. The issue here is that we need different models to explain the general situation from the specific problem which an individual agent faces.

Collective action problems do not of course exhaust the account of power. Agents have resources that they utilize in order to dominate others. Again, we can examine the specific resources of any token agent, and we can think about the resources that types have in relationship to other types of agents. When we think about the power and privilege structure in general it is to this type-level that we turn.

I've not said much about my distinction between luck and power, especially systematic luck and power, in this chapter. I think it is expressed well enough in *RCPP*. It is the most critiqued aspect of my account however (Barry, 2002, 2003; Lukes and Haglund, 2005; Hindmoor and McGeechan, 2013), but I have answered those criticisms in other places (Dowding, 2003) and do not want to repeat myself here. I do think that a large part of that critique is normative, saying a type is systematically lucky seems to let them off the hook. Perhaps had I used the term 'privileged' to describe some of those who are systematically lucky, as I have repeatedly here, the criticisms would not have such heat. To be sure, many people who are systematically lucky are *also* powerful, but being lucky they often do not have to exercise their power. One cannot just assume because a group is advantaged that they are exercising power, nor that they have power they are not exercising. And structurally attacking privilege will require different tactics from structurally attacking power. That is why I think my distinction is an important one.

The issue of reputation is related to systematic luck and power. My use of it is a little problematic. What I have tried to do in this chapter, which I do not make clear in *RCPP*, is that reputation is really important in the dynamic analysis of token power games. Without it, we will not understand the tactics of the players. However, it is very difficult to study in comparative static analysis. Not impossible, the early studies of community power looked at reputation – not quite in the same manner in which I use that concept – but one can interview and survey people to gain insights into how they think other agents will act and respond to their actions. That will give insights into the reputations of different types of people.

10

The Nature of the Exercise

10.1 Conceptual analysis

The subject of *RCPP* is political power. In that sense the book is an exercise in conceptual analysis. Unusually, perhaps, for such an exercise, I never actually define power. I do provide a semi-formal definition of 'power over' that I term 'social power', and similarly provide a definition of 'power to' that I call 'outcome power' (*RCPP*, p 48), but those definitions are used for a specific purpose: to suggest that logically 'power over' is a subset of 'power to'. I provide those semi-formal definitions as part of an argument that 'power over' is a subset of 'power to', a conclusion obvious from the definitions. My intention of providing the semi-formal definitions is to challenge those who insist 'power to' and 'power over' are logically separate to state explicitly what is wrong with my definitions of those terms. I do not provide a formal definition of power overall, since I think generally that providing such definitions are otiose. I am trying to analyse power, and I do so by a method that suggests how we could empirically measure power, even if only roughly.

Of course, I have no problem with authors defining key terms if they do so for the sake of clarity in a particular article or book. It is fine to state that one is intending to use a concept in a particular way and provide a definition for its use, especially if other writers use that term differently. Such definitions are useful to ensure that there is no ambiguity in the usage; I take it that they are made for the sake of clarity in that particular argument and are not supposed to be extravagant claims that this specific usage is what the term 'actually means'. Defining a term for a specific purpose is often a good idea, but one should not think that that purpose is the only use to which a term can be applied. Indeed, it seems to me that one of the problems of analytic political philosophy has been the constant search for necessary and sufficient conditions for the correct application of such terms as 'political power', 'liberty', 'coercion', 'welfare' and so on. Too many articles seem to be stand-alone pieces attempting to define a term for all times and places.[1]

Of course, when one does engage in conceptual analysis, it is hard not think in terms of providing necessary and sufficient conditions for the correct application of a term. However, if a language is not logically closed – and indeed is developing, due to our changing knowledge of the universe and our evolving morality and social conventions – there is no particular reason why we should think we ought to be able to provide necessary and sufficient conditions for all applications of a given term. For fully formalized languages where we expect proofs, we can demand necessary and sufficient conditions. That is the nature of proofs. Philosophers, who tend to think a counterexample suggests a problem for a definition of a term, meaning one has to rethink the concept, seem to believe that they are engaged in an exercise analogous to mathematics. Finding a counterexample is akin to having destroyed some kind of proof. In fact, it seems to me, counterexamples deserve no more attention than we would expect, given the intrinsic interest of the example. That is to say, the more unusual or bizarre the example the less interest it has for social theory. Philosophy is not mathematics. Counterexamples should be treated in the same way that outliers are treated in empirical social science, as something to be explained beyond the analysis that suffices for most cases. Furthermore, social theory should recognize that it is engaged in the analysis of a moving target and so we should expect conceptual innovation.

The best analyses of social concepts are not simply attempts to provide logically coherent definitions for all times and places (for example, Oppenheim, 1981), but rather tie the concept to a specific normative theory – of justice, freedom, the state. For example, Pettit's (1997, 2012) conceptualization of liberty as freedom from domination is, at least in part, designed to defend his preferred account of republicanism against liberal conceptions of the just state. The best analyses can be seen, then, as providing a definition of a concept that coheres with other concepts within a grand normative theory. In that sense, the concept is tied to a specific task. If that task is expressly normative, the conceptual analysis is akin to an exercise in public relations. If it is essentially scientifically predictive, then it fulfils a role that is useful for explanation. Often in social science, conceptual analysis straddles the explanatory and public relations exercises. This is especially true since specificity of concepts within the grand normative theory is often somewhat lost, because theorists also want their grand theory to be *the* correct one, and one of the ways in which they press their case is through the conceptual analysis of the terms that enter into the argument for their theory. They try to show that their version of the concept best fits our considered judgements over the use of the term.

Terms are not given necessary and sufficient conditions just for their logical coherence within a theory, but also because they fit, to a greater or lesser extent, with how that term is used in ordinary language. Appeals are made explicitly or implicitly to our intuitions about the concept. We are asked to reflect upon how we use the concept, how it fits with other concepts and our beliefs about the world, to come to a considered judgement about its correct application. When it comes to normative theories – about social justice, or what the free society should look like – our moral intuitions are already implicated in how we view the concept. So conceptual analysis is not simply about logical coherence, but also coherence with our established moral views. Of course, our views can change with argument and considered judgement, and some of that argument and judgement can be over the rival definitions of, say liberty, as to what precisely those rival definitions entail. Such disputes over definition will be 'merely verbal' (Chalmers, 2011), even though the underlying moral argument will not be.

At the end of the day, conceptual analysis over normative terms will always be contestable, given a plurality of moral views. That is the basis of the thesis of essential contestability – at least in its 'value plural' version (MacIntyre, 1973; Gray, 1977; Lukes, 1977, ch. 8; 1991, ch. 3; Gerring, 1999, 385).[2] In Chapter 8 of *RCPP*, I provide a version of the 'subscript gambit' discussed by Dave Chalmers (2011) to overcome verbal dispute. I suggest that such value-plural contestability is not the deep problem Lukes and others claim. Different uses of the term 'interest' within different theories can simply be subscripted, to allow us to concentrate upon what is importantly different in the theories themselves. I think the subscript gambit can overcome *conceptual* dispute across different accounts. It does not, of course, solve our problems about what we should do given we have different value systems.

Different people using the same word to mean different things is a form of ambiguity; the subscript gambit can clear up the confusion that may ensue. There is, to be sure, fundamental disagreement about how we conceptually partition the universe, but as long as we are aware of the partitioning of our antagonist there is no conceptual dispute. The disagreement might be over the best way of partitioning. In science, the best partitions are those that provide what Daniel Dennett calls 'real patterns' and thence give the best predictions. In normative dispute, the disagreement is a moral one about how the world should be. It should not be a disagreement about what words we use to describe how the world should be. To be sure, some words have, in ordinary language, particular appraisive force. So getting the community to use a given word in a way that supports your theory allows that appraisive force to spill over into your preferred theory to your advantage (Bosworth and Dowding, 2019). That is why I describe some conceptual analysis as a public relations exercise.

I can know what you mean by a term, but may dislike your usage of it, and vice versa. Only if there is an original exemplar – as Gallie argues with regard to Christianity and the word of Christ or Marxism and the word of Karl Marx – one can argue that the concept is essentially contested, because then the dispute is over which extension best represents the exemplar. However, given developments in our social and moral life, there might be no answer that can be deemed correct by an external determinant (Evnine, 2014). Grafstein (1988) argues that genuine essential contestability exists only over real concepts – ones that refer to real items in the world. Perhaps it is possible to have two (or more) rival ways of partitioning the universe, each giving multiple concepts, that are similar but not identical. However, again I think the subscript gambit could prevent dispute over the concepts themselves. Moral objectivists might think that we can have conceptual dispute over what the real moral objects are. But I take it that moral objectivists would then think that only one side is correct, even if we have no way of demonstrating that fact. Moral realists, who I take to believe that there can be several equally real moral systems (in the same way that different societies might have different conventions to solve similar social problems), might also engage in conceptual dispute, but again the subscript gambit ought to suffice.

However, this sanguine view of value-plural essential contestability does not mean that the news about conceptual analysis in social theory is all good. I believe that many moral and political concepts have a *deeper* problem. The problem is not so much that different people hold different value systems, but rather that, for many complex social terms, we *each* hold conflicting intuitions or beliefs about what they entail. We find that when we interrogate and analyse such complex concepts carefully, they turn out to be inconsistent with some of our cherished beliefs about them. This fact of incoherence is hidden by the fact that they are vague (Dowding and Bosworth, 2018). It is only when we precisify them that we discover this incoherence.

I think we uncover this incoherence when we stop verbalizing our dispute and shift to the language of science – that is, to the formal languages of logic and mathematics. When we come to measure what our concepts entail, any incoherence emerges. We can see this in terms such as 'the collective will' (Arrow, 1963; Riker, 1982), 'freedom' (Dowding and van Hees, 2009) and 'democracy' as we try to think of ways of measuring them. Where we find incoherence, we have two alternative strategies. Which should be chosen depends upon the purpose for which the concept is being used. First, we make 'coding decisions' – a form of the subscript gambit – in the knowledge that our coding is subject to (at least marginal) error. For some concepts and research questions this is the best ploy. We can see some aspects of conceptual analysis in empirical social science as coding decisions, so we should not be too vexed if some researchers conceptualize a term differently from others. Such coding strategies are commonplace in empirical social science where qualitative concepts need to be carefully coded in order to allow quantitative manipulation and analysis. The issues and problems with such coding decisions are well known amongst empirical researchers and I think many philosophers engaging in conceptual analysis for specific purposes could learn from that literature.

The second strategy, which is more appropriate when we are engaged in high-flown conceptual analysis for grand theory, is more radical. When the incoherence is at the heart of the concept, the term itself needs to be eliminated from scientific or academic analysis. Elimination is not simply the same as specifying the necessary and sufficient descriptions of a term's referent and then using the description rather than the term (Lewis, 1970). That would simply be a long-winded way of saying the same thing. Rather, where there is disagreement, we reduce the term to those component parts which seem to jointly provide confusion in order to focus upon the problem. If those parts are in themselves coherent, then we have precisified in a manner that genuinely represents the world. If we have not, they still do not refer and we have to eliminate again (Bosworth, 2016; or Chalmers, 2012 for a more general account). We find then that, for terms such as 'democracy' and 'liberty', some of those things we hold as being most valuable and which are jointly necessary for the vague concept come into conflict. Sometimes the conflict concerns a weighting problem. We have different elements within the concept of democracy, such as voting rights, freedom of speech, independence of the judiciary; and how we measure and then weight those different elements is problematic. In fact, weighting for elements within the concept of democracy is a well-known coding problem. Sometimes, however, the problem is deeper: it is not just a weighting problem. For example, we find that different ways of measuring the amount of freedom of choice – be that in the number of alternatives in an opportunity set, their diversity or their intrinsic value – will dominate the measure. Any way of so measuring bumps up against our intuitions about what freedom means. What we then need to do is eliminate the term, replace it with terms that precisely mark the issues we consider important, and measure these separately. That does not mean we have to banish terms such as 'liberty' or 'democracy' from our vocabulary altogether – that would be ridiculous. However, we can recognize that we always use such terms in a vague manner. Often this will not matter, where it does, we have to eliminate and then ensure that our claims are based on coherent components of the overall incoherent term.

It is for these reasons that I am not too concerned about providing formal definitions for words, and only do so for clarity in the specific context of my argument. It is also why I believe that the real test for the coherent use of a concept is how, at least in theory, we measure it. I realize that many people will be aghast at that claim, but my mantra 'if you can't measure it, then it doesn't exist' is made remembering that the first order of measurement is nominal (Krantz et al., 1971). If you can give something a proper name, then you must be able to distinguish it from the rest of world. Once you can do that, you can start counting it, partitioning parts of it, and so on. Students sometimes ask 'Can love really be measured?' As soon as we make statements such as 'John loves Mary more than Mary loves John', we are making some claim about quantity, which begs some kind of measurement exercise. We might ask 'in what way does John love Mary more than she loves him?' and that question demands some criteria for justification: that is, some form of criteria that can be roughly measured. I think we can measure all sorts of things that are part of our intuitions or beliefs about democracy, or about freedom or about collective will. But we struggle to find something that measures all those elements that do not fundamentally conflict in one way or another.

These complex but vague terms can be thought of as latent variables (Treier and Jackman, 2008), theoretical terms that reduce the complexity of their elements; but how we choose to combine or weight those elements will affect our measurements of the concept. Our problem is that for complex normative terms we have different intuitions about the way to measure and weight the respective elements. For vague political concepts, for scientific or academic analysis, we need to eliminate the term for the purposes of analysis. The term is not eliminated from our natural or folk language, but we understand that our natural language is vague. We only use the eliminated terms when we explain generally what our academic analysis has shown.

In the natural sciences, precisifying terms often produces very different definitions from natural or folk usage, but the new theoretical term helps explain the phenomenon captured by the folk usage. For example, the scientific definition of 'heat' is, roughly speaking, 'the transfer through contact of thermal energy between two systems at different temperatures'. Our manifest feeling of heat comes about when we come into contact with a system that has a higher temperature than we do. The folk understanding of heat can be explained by the scientific one. Ideally in social science we should want to develop concepts that also underlie and explain our folk understanding of manifest phenomena that we come across. When I engage in conceptual analysis that is what I am trying to do. The concept might depart from our folk understanding, but should help to explain why we have that folk understanding of the manifest phenomena we witness.

Thus, in *RCPP* I am trying to provide an account of how to analyse power in society from which we can derive many of our folk beliefs about its manifestation. The distinctions I make – for example, between power and luck, specifically power and systematic luck – are designed to show that the best way of analysing manifest aspects of the power structure require such distinctions. I have to say that most (though not all) of the critiques of the concept of systematic luck seem to object to the word 'luck' rather than the distinctions I draw. I think critics drew the implication that privileged people are 'off the hook' because it is their privileged status and not their actions that advantage them. To some extent, I do think it lets them off the hook, and rightly so. In my view you cannot blame someone personally for being privileged. That is not to say, however, that we should not be morally outraged by such privilege. We can blame the system that makes some people privileged. To my mind that is precisely where blame should be laid. Furthermore, changing the social structure that privileges some requires different types of actions than those needed to get privileged people to act differently. (We can pursue both, of course.) The fact that privileged people are often powerful as well is beside the point. We can blame them for exercising their power and blame the structure for making them systematically lucky. We can also blame the system for making some people systematically unlucky.

RCPP was not written in order to provide a new understanding of power because I disagreed with earlier definitions. It came about because I felt a specific concept, the collective action problem, could explain some of the phenomena that Lukes and others grappled with. Given that factor, and using decision- and game-theoretic work on power, I thought we could come up with a concept that is measurable (at least in theory). If power is a complex and vague term akin to democracy, liberty or the collective will, then we might have to eliminate it in favour of some other concepts that form part of our folk understanding of power. To some extent, breaking power down into the five elements I outline is precisely that kind of elimination or reduction.

I do not think, however, that power ultimately is an incoherent concept like liberty. To be sure, 'being powerful' is vague in the same way that 'being rich' is vague. But being rich is not complex in the manner of democracy or liberty. We can measure relative wealth, and we can also measure relative disposable income (to make sense of being 'house poor'), so we can precisify relative wealth and disposable income. Such measurement exercises are relatively easy. That is not to say they are straightforward. They require in practice that we take into account all sorts of factors, compounded if we are comparing across different countries or cultures. Nevertheless, these are precisely the sorts of measurement problems that empirical social science grapples with all the time. With power, we can measure the de facto and de jure authority that different agents have, not precisely, but as well as we can for many social science concepts. Who controls information is also difficult to measure, but we can gain some insights into who controls information flows and how many people are affected by them. We are also learning the importance of framing effects.

Measurement is not easy, but it is the stuff of social science. Weighting these elements might mean that coming up with overall power measures for given agents comparatively will be contestable. However, the issues we are more concerned with – the distribution of power within a society – is more readily discernible using these sorts of measures. Finally, there is an external check upon our measurement claims: namely, who in the end gets what and how. As long as we take luck into account, we can make judgements about power; and with luck we can make judgements about the power and the privilege structure.

I think, then, we can have a rough measure of the relative power of agents in terms of relative resources. Yes, 'power' is a vague predicate, like 'rich' and 'tall', but, like these terms too, it is not incoherent. We might have some difficulty in weighting its elements, but given we are considering outcomes, we do have some way of assessing different weighting schemes. When it comes to more appraisive terms such as liberty or democracy, we lack such obvious outcomes to provide the check on our claims.

Why do I think that power is not like freedom or democracy? Two reasons. First because I think power is not as normative a concept as freedom or democracy, or even 'the collective will'. Whilst freedom, democracy and the value of the collective will not always have been identified positively, generally speaking and certainly under liberal understandings, they are identified as positive virtues. Power, and the structure of power, is not. The power to do great good can also be thought of as the power to do great harm. When we talk about 'the power structure', we are talking comparatively. We are generally talking about the description of power-holding with the underlying assumption or belief that some power structures – perhaps more egalitarian ones – are preferable to other forms. In other words, power seems to be more descriptive and less normative than the other terms. This relates to the second reason why I do not think power is a vague concept. We have a fairly good idea of what power means outside of social contexts. We can talk of the power of engines, of animals, of the wind or tides, and so on. Power in science has a fairly straightforward meaning that can be traced to the energy (the resources) that the object possesses. So my account of power transfers that more general scientific understanding into a social context.

I need to mention a caveat about the non-normative nature of power. In my view, all concepts, even scientific ones, are normative to a degree (Dowding, 2012, 2016a). That degree might only be the extent by which we want to conceptualize the universe, the manner in which we want to partition it into different categories and items. We do so for a purpose, and creatures with very different interests, or perceptual apparatus, might want to partition the universe differently. Real partitions enable scientific predictions, and any creatures that rely upon such predictions, or search for them, will converge on the real partitions – so their concepts will also converge. But, of course, not all creatures do need or search for many partitions. When it comes to communal life, some of the most important partitions are social ones, and conventions, expectations and norms vary across human societies. However, these partitions must be social equilibriums, and so we can rule out some sorts of behaviours as being long-run conventions. Whilst societies might not converge totally, they cannot completely diverge.

This is, of course, a way of reintroducing value-plural essential contestability – but here we can use the subscript gambit to ensure that our dispute remains in the normative realm over how we want to make moral judgements and not over concepts themselves. We can each mark concepts by subscripting any given term by which we want to label those different concepts. If, for example, someone insists that 'political power' is normative, because any use of political power is always wrong, then they will have to defend that use by the way in which they define politics and power. I can simply say that is not what I mean by the term 'power'. If we both agree that doing x is not wrong, but I want to say i doing x is a use of i's power, and my antagonist cannot say that since she has defined power as being wrong, we agree about the phenomenon, but not on a word to label it. This is then a verbal dispute (Chalmers, 2011) akin to ambiguity, not vagueness.

It is for these reasons that I think we should always try to define concepts in as non-normative a manner as possible. We should make our terms as value-free as possible so as not to gerrymander our arguments to our preferred conclusion. We should also make them as value-neutral as possible, in the sense that they are not defined in order to defend one grand moral theory over another (Dowding, 2016b, 194–8). Making them as value-neutral as possible enables discussion with those who hold different normative commitments and beliefs.

10.2 Type and token

The distinction between type and token is important in my analysis of power as I tried to make clear in Chapter 9. In fact, I think it is a key distinction in social science generally. A type is a given class that is constituted of many token examples. A token is an example from a given class (Wetzel, 2009; Dowding, 2016a). A token is a member of type given some of its characteristics; any single item can be a member of many types. Analysis of types and analysis of tokens belong to separate sets of questions, though answers to type questions are implicated in answers to questions involving tokens. I see social science generally providing explanations in terms of mechanisms (Hedstrom and Ylikoski, 2010; Waldner, 2012). Both the collective action problem and solutions to it are mechanisms. Mechanisms provide the structure, or the structural causal story, and these stories are about types. Mechanisms are explanatory of types and are explanatory of any given token to the extent that the mechanism applies to that token. And the extent to which it applies to any given token can vary. I believe one of the biggest causes of confusion in social science is conflating explanations of types with those of tokens (Dowding, 2016a; forthcoming). I think this confusion extends to the analysis of power, as I argue below.

I discussed conceptual analysis in 10.1 above, but did not say what a concept is. I take it that when we do conceptual analysis we are attempting to describe a type. That is, we are characterizing a type by the features that identify tokens as members of the type. Now, we might note that a type is not necessarily described by characteristics common to all its token members (Wetzel, 2009). For example, there might be no features that are shared by each and every member of a species and which is a defining feature of the species as a type. Indeed, as Wetzel argues, that fact follows as a result of Darwinian fitness. Types are used to help us explain type-level phenomena. Any given token story might depart in important ways from the general mechanism of the type-level explanation. So the full explanation of token-level phenomena might go well beyond the type-level characterization.

Types might be the subject of explanation – in statistical language, the dependent variable – or they might be part of the explicans – an independent variable. My analysis of power is, of course, of power as a type. When we look at actual power games, we are giving token-level explanations. The type-level analysis should provide the background for the token-level explanation, but the token level is likely to mention aspects that go beyond any type-level analysis of power relations. Where we see power as a capacity, for example, we are considering what processes could occur. When we analyse an actual power game, we analyse what did occur. To explain what did occur rather than what could have occurred, we might investigate aspects that go beyond our type-level explanation – for example, what the agents' preferences were and why they had those preferences. The latter question might invoke other type-level analyses.

As I have suggested, I see little point in trying, from first principles so to speak, to conceptualize social phenomena. Rather, we need to conceptualize those types that contribute to our best theoretical explanation of the type-level phenomena we witness. My aim in *RCPP* is to try to understand the nature of the power structure – why some groups dominate others – and why that structure of domination persists over time. I think everyone would agree that some elements of why some dominate others is obvious. Some people have more money, greater physical strength, more education, more friends, more guns, more property or other physical attributes – in short, they have more resources. This enables them to physically dominate others, to threaten or cajole. Some people have laws and regulations on their side, giving them the backing of the state to dominate others. That gives them some measure of luck, and resources they can use to their advantage. However, beyond those obvious physical manifestations of social and political power is the puzzle of why the weak do not band together to fight back, why they seem to acquiesce willingly in their own subjection.

I give the example of the overthrow of the Romanian president Nicolae Ceausescu. He was overthrown on a particular day, but could not the mass have removed him the day before that one, or the day before that? The answer is probably yes; but the precise day it occurred is a contingent matter concerning token events. Our more general type-level query, on such an event, concerns the mechanisms by which revolutions occur. The collective action problem provides us with a mechanism to explain why revolts do not occur spontaneously when they would be in majority interests, and in the interests of those who, collectively, have the power to revolt. Analysis of revolutions, of course, gives us evidence on the different mechanisms by which collective action and mobilization occur. So I think the collective action problem can explain a large part of the lack of power of individuals; but in making my case, I also need to distinguish between power and luck. That distinction helps answer the puzzle of why some groups – in *RCPP* the example was British farmers – get what they want more often than some other groups, in ways that seem beyond their resources. The farmers were protected by governments because of their perceived strategic importance, enshrined in promises made early in the Second World War. Their privileged position was maintained, despite the waning of their strategic importance.

RCPP is a book that has an argument concerning the best way to analyse power in society. It is directed at type-level analysis. Whilst it offers a general account of how we analyse power, it is supposed to provide guidance about how we can actually measure power. In the first instance, in a comparative statics sense, we can analyse the different power of types of actors within any society by the resources they command. We can also compare the relative power of types of agents across different societies, again by the resources they command. Such an approach ought to allow us to measure the relative power of presidents at different times in a political system or the relative power of presidents in different political systems. We can do the same for social types. We can look at power disparity across the sexes, for example institutionally by looking at employment laws, laws governing the household, and social laws more generally. We can also examine the actual material holdings of different classes of men and women, and the educational capital that each group commands.

Power also depends upon the exercise of one's resources. How one uses one's resources depends not just on how strategic and lucky one is in the game of life, but also on one's expectations. These effects on power relationships are harder, but not impossible, to measure. For example, we can gain an insight into actual discrimination by using John Roemer's Equality of Opportunity (EoQ) measure (Roemer, 1998, 2002). Within any given type – men and women, ethnic groups – there will differential outcomes. Within each type, some will earn more, live longer or have better health. If we assume that 'talent' is equally distributed across the types, then by comparing across types at each centile within the in-type distribution, we can gain an insight into the discrimination or advantage that that type gains. The differences within a type ought to reflect the differences in talent and in luck that these agents have had relative to others in their type. Differences in outcomes across types will reveal differences in the relative success of the different types. So Roemer's EoQ measures success of different types. What is to be explained is how much that differential success is due to systematic luck and how much to power. Either way, it is due to the power and privilege structure. Public policy needs to be informed by this kind of measure: the statistical distribution of outcomes across types of people, and across different social and political systems. And it needs, in my view, to acknowledge the effects of both luck and power.

So *RCPP* is an exercise in conceptual analysis, but of a term that I take to be capable of being conceptualized precisely because it labels an empirical phenomenon: the capacity of actors. My analysis is designed to suggest ways in which we can measure that capacity. The capacity itself is not especially normative, but the structural differences in its levels and distribution are. That is, we can disagree over the socially optimal distribution of capacities, or whether exercising power in one way or another is legitimate. Nevertheless, I see disagreement over the concept of power as being largely over the disambiguation of different uses, which can all be represented by a resource-based account.

Some will view my claims very sceptically. They might say: even if we accept your claim with regard to the most obvious aspects of power inequalities in terms of all sorts of material resources, the real issue concerning social power is the behind-the-scenes stuff. The real issues concern domination and how people accept it. They concern not what agents do given their preferences, but how preferences and interests are formed. They concern the very identity of agents. Your rational choice analysis takes agents as given and it misses those aspects that Lukes discusses in the third dimension of power. As I discussed in 9.2 – the answer is that the model that looks at the token individual takes their preferences as given, but the model that looks at why people like them – their type – have those preferences explains where they come from.

Models of type can help us to understand the structural determination (or what I call structural suggestion) of individuals' preferences. I will discuss the relationship between structure and agency in 10.3.

10.3 Agency, identity and structure

RCPP argues that power is a capacity of agents. The power that agents have is a causal power, the ability to do things. Of course, agents operate within a social system, so they are not always able to achieve their aims because of what other agents do. Agential power is determined not only by one's resources and one's decision to use those resources in a given manner, but also by others' resources and how they choose to use them. Agents can get far more done collectively than they can individually. And agents can face contrary forces if other agents choose to oppose them. That is what makes individual power a form of social power, since it operates within a social context.

We can measure this social power in a static or a dynamic manner. A static manner compares the resources of different agents in order to judge their comparative powers. It maps the structure of power by resources.

Of course, agents can choose to use their powers or not. That choice, in some contexts, can be considered contingent. So how well one does in any given situation, whether that requires the cooperation of others or is a strategic battle against them, might depend upon some measure of luck or chance. In some contexts, the choice to use or not use one's resources might be considered to be part of the power structure. Collective action often depends upon expectations about others' behaviour – it relies on the institutions and norms of society. Two groups in different societies might have equal material powers; in one sense they have equal powers to achieve some collective end, but, practically speaking, they might not be able to deploy those powers. In one society the institutional framework may enable collective action, whereas the norms and expectations of the other society may hamper it.

Some writers spurn the type of account of power that I offer because it is agency-orientated. It seems to privilege agency over structure. I naively believed the agency–structure debate was already dead by 1991, since it is obvious to me that structures are composed of all the actions of agents, and structures provide the incentives for those actions. One cannot study one without the other. It is true that we tend to privilege one over the other in certain types of explanation. We concentrate on structural explanations when we discuss types, and we tend to give agency explanations when we explain token processes. Since *RCPP* is about very general ways of examining power in society, I see it as providing structural explanation, notwithstanding the claim that power is a capacity of agents. That is because, despite the odd casual example, all of the agents in *RCPP* are types. In other words, they are defined structurally.

Here I want to concentrate upon a more general worry about how agential accounts ignore structure. In *RCPP* I view structure as the relationship between variables, where the variables in this case are agents. That seems to imply that agents are completely separate from their relationships, as we might see dots being different from the lines constructing them – which is one way in which we might view a social structure. However, structures go deep inside each individual agent, since the structures around people affect their very expectations, attitudes and interests. I have called the first view of structure 'surface structure' and the second 'deep structure' (Dowding, 2008). *RCPP* says little about deep structure other than pointing out that we can make a distinction between endogenous and exogenous interests, and showing how we can model objective interests.

Deep structure suggests that not only does our environment create incentives for people to behave in one way or another given their preferences, it also creates those preferences. One way of thinking is to say that individual i believes x, desires y and, given structure Z, acts in a manner to achieve their aim. This is the standard economic or revealed preference manner of thinking about human behaviour, and the one adopted in *RCPP*. A rival way is to say that structure Z creates my belief x and desire y. Here structure becomes the agency, as it creates the neurological activity that causes human beings to act as they do. As John Roemer (1986b) shows, we can think of historical processes as structures ('solution processes') leading to preferences or as preferences leading to structures ('outcome processes'). If structures fully determine preferences, then we ought to be able to tell history without mentioning human preferences, interests or beliefs. If structures are fully determined by actions and actions are the result of fully autonomous beliefs and interests, then we can tell history just in terms of what agents think they are doing. I guess some people might think that the first is plausible; some think that the second is plausible. However, I know of no serious attempts to recount history either way round.

History by preference alone would be implausible, because the very identity of people – how they view themselves, their interests and their beliefs – is, at least in part, determined by the environment around them. J.S. Mill was a radical liberal for his time, but he was also a supporter of British colonialism. His attitude to the British empire is definitely a product of his time. Indeed, it is difficult to imagine any British person in his time having the attitudes towards colonialism that are commonplace now. But it is equally difficult to imagine anyone today with Mill's attitudes on freedom, equality, the law, feminism, having anything remotely like his attitudes to colonialism. What is reasonable to believe in one time and place is very different from what is reasonable to believe at other times and places (Dowding, 2013, 73–9). If you disagree with that claim, then you are taking 'reason' out of history and out of all structural influence.

Nevertheless, it is equally implausible to think that Mill's beliefs and interests were fully determined by his position in society. After all, he had some unique beliefs. We do not need to claim that in some sense his beliefs were not physically determined: perhaps it was his circumstances combined with the specific ways the proteins were expressed in his genes together determining the precise ways his neurons fired. Or maybe we prefer a story that gives him autonomy in some more Kantian fashion. Either way, it is something about *him*, and not just the structure, that jointly determines his beliefs, desires and actions.

For any historical story we need to bring in both the agents and the environment that creates their incentives to act in one way or another. That is why I like to write of 'structural suggestion' rather than 'structural determination'. Structures suggest reasonable ways of thinking, believing and acting, and agents choose with those suggestions in mind. What we choose to concentrate upon – agency or structure – when telling any historical narrative, or what we choose to concentrate upon in any explanation of social institutions or social outcomes, depends on the question being asked. We can utilize a direct analogy with the old chestnut of genes and environment to explain this.

The old question 'nature or nurture?' is, to a large extent, a stupid one. For any outcome it is always both. However, what we choose to focus on – genes or environment – depends on the question we ask. Take two sheep, Dolly and Dilly. Imagine both are clones and so are genetically identical.[3] Bring them up in different environments – feed them different foods at different times, keep them at different temperatures, and so on – and any difference in their behaviour will be fully determined by their environment. Take two sheep, Dally and Dully, that are genetically different and bring them up in precisely the same environment: any difference in their behaviour will be fully determined by their genetic differences. But for all four sheep, their actual behaviour is determined by both genes and environment. 'Nature or nurture?' only makes sense as a *comparative* question – not about the individual token sheep, but about differences in the behaviour of *types* of sheep. Is one breed more aggressive than another just because of their genome, or it is also due to the environment in which they live? We could find out with the appropriate experimental set-up. Nature versus nurture debates are comparative questions about types, and only through types do they have an impact on token examples.

The same is true of structure and agents. We do not need to think about human agents' behaviour as being genetically determined; we only need to think of them as being autonomous – that is, there is something in themselves that leads to their behaviour given their environment. That environment includes the actions of others and the autonomous agent's expectations about the actions of others. The greater the similarity in the environmental conditions which agents experience, the more we will think autonomous actions explain differences in their behaviour; the more alike the agents behave, the more likely we are to think that structure is where we will find our explanation. The explanatory issue is where do we expect to find the greatest variance for our question? When we are looking at types of individuals, we are obviously looking at structural differences. First, because differences in the behaviour of the same type will be explained by differences in the incentives those types face in their different situations. But second, where we are looking at types in the same situation, we are defining the type by their structural characteristics.

When we look at real, token, individuals, however, we are seeking the nuances of their behaviour. We do not deny the structural influences; indeed, we might conclude in our narrative that the agent had no choice but to do as they did. Often in history we pick out extraordinary individuals precisely because they took actions that were brave, heroic, unexpected or brilliant. They did things, whether they be intellectual or strategic, that could not be expected. What we are doing, then, is comparing those individual actions counterfactually with others who might find themselves in that situation. The upshot is that the agency–structure issue is always a comparative one.

When it comes to issues of power, however, the critique is that we have to delve into the interests of the agent themselves in order to see how the power structure affects them. I considered this in Section 9.2. I will just add here that if we attribute all causal power to deep structure, then we leave no room for free will and autonomy. Few would be willing to accept this (Dowding, 2008). It is surely correct that our beliefs and our values are generated by the physical and social environment around us, but also correct that we can reflect on those beliefs and values. It is that reflection that enables humans to criticize, to create new beliefs and value systems.

The issue of deep structure is particularly important to feminist accounts of power in both negative and positive ways. And the issue of agential autonomy cuts through feminist discourse on the power structure. Allen's (1998) discussion of two major ways in which feminists have discussed power illustrates this issue. 'Empowerment' theorists concentrate upon the collective action potential of women and how women can learn to empower themselves. 'Domination' theorists concentrate upon how men dominate women. This obviously includes the ways in which men can explicitly and consciously discriminate against, use and abuse women, but also explores how women acquiesce in their domination. And it is this acquiescence that is part of the deep structure. Domination theorists claim that the difference between men and women, between masculinity and femininity, is itself part of domination. If we place too much emphasis on the structural elements that lead to acquiescence, we turn people into automatons that can only respond to incentives. If we place too much emphasis on their taking responsibility for their actions, we end up blaming the victims.

The only way out of that kind of dilemma is with a two-fold strategy. The moral strategy: people need to take responsibility and make hard decisions sometimes. The political strategy: we need to change the incentive structures, first by changing our laws and regulations, and second by transforming people's expectations. That second element of the political strategy cannot be achieved without the moral strategy working.

Whilst the moral analysis and strategy is important, in my view political philosophers have concentrated too much on it and not enough on the political strategy. There is too little prescriptive work on how we change the rules to help people overcome their collective action problems. We need both analysis of domination and analysis of empowerment, and we need strategies to connect the two.

10.4 Summing up

RCPP has been published in its original form. If I wrote it now it would be a very different book, and much longer. However, I still doubt that its lessons have been learned by enough people who work in power studies. I am convinced that examining agents' resources is the key element to understanding the power structure. I am convinced that we need to see how types of people operate, why they behave as they do, to fully understand the workings of the power and privilege structure, and I am convinced that the collective action problem is the key to understanding that structure. There is more to say, of course, but merely stating that we need to look at where preferences come from is not enough: we actually need to study them. And that in part is an empirical job. The theoretical element comes in modelling how we can expect attitudes to change with changing forms in society.

Preferences are formed in part by the information we receive, and there are plenty of studies of how issues are framed and biases formed in the massed ranks of the media. The normative question is how much of that bias is manipulation – the use of power – and how much unconscious. I addressed these issues in Dowding, 2016a and 2018. Those articles, as much as the argument of *RCPP*, show how important conscious intention in action is to how we judge power. It is also important for the power/luck distinction. Yet when we look at resources in comparative statics analysis, we take no account of intention. We are only looking at capabilities. And we have to consider capabilities only in comparative statics or we commit the exercise fallacy (Morriss, 2002). When we examine actual power dynamics, however, we cannot avoid intention since we have to interpret people's actions, and that requires intentions. So again, we have to be aware of the questions we ask and the type of analysis we are conducting when we think about what goes into our study of power.

There will always be more to say on the subject of political and social power; but whatever is said, it ought to help us to recognize when power is in use, and it ought to be able to help us transform the power and privilege structure into a better form of society.

Endnotes

Chapter 1: Introduction

[1] In fairness to Giddens, it should be pointed out that he uses the term 'structure' rather differently from the use here. Giddens (1984, 25) suggests that structures are 'recursively organized sets of rules and resources ... out of time and space, save in instantiations and co-ordination as memory traces, and ... marked by an "absence of the subject"'. The actions of people 'recursively implicate' the structure. It seems that structures here are closer to what most of us call norms, conventions, rules and so on, though none of these are out of time and space. However here I do not have enough of the latter to consider the relationship between Giddens and what I say.

[2] Note that this does not suggest that non-acting objects like sunspot activity or the behaviour of a set of computers may not be major causes of social outcomes. These are events and so can be causes of other events.

[3] What constitutes an action – which rules out, say, what present-day computers and robots do – is discussed in Chapter 2.

[4] The colour case, a favourite example for explaining supervenience, is different from the other two. Here the same property of the surface, viz. reflecting the red wavelengths, causes us to see 'red' in all cases. But the surface properties of some objects which would reflect white wavelengths cannot do so when there is no white light present.

[5] This is an assertion for which I offer no defence here. I would use some Wittgensteinian arguments in any attempt to do so.

[6] We may speak of organizations or groups acting, but they only do so through the actions of individuals. For some explanations of why organizations acted as they do this reduction is very important. We may talk of a computer or a robot acting, but it is only mediating the action of the individual who programmed it: it is the programmer who had the beliefs and desires and not the computer. This is so even if the results of the computer's actions are not those intended by the programmer.

Chapter 2: Rational Choice and a Theory of Action

[1] Some people talk to psychoanalysts or priests, the rest of us to our close friends, because we believe that others may help us to understand ourselves better than we can on our own since they have a different perspective.

[2] I believe that Schelling coined the phrase 'bettering oneself to death' but I have lost my notes as to where. Dummett discusses the problem (1984, 33–5), where he also gives an example of apparently rational intransitivity.

3   This is a slight simplification. There are three sorts of behaviourism: ontological, methodological and analytical. The first maintains that only 'the behaviour of organisms' actually exists and not consciousness (Watson, 1930). The second holds that, even if consciousness does exist, a true science can only study behaviour (Skinner, 1953). The third claims that psychological concepts must be susceptible to examination through publicly accessible acts (Ryle, 1949). The theses become less absurd from one to three. The third in its epistemological form (Wittgenstein, 1953) is compatible with my argument.

4   It is interesting to wonder why certain decisions should ever have been called 'non-decisions'. I think it may come back to the systems methodology of Easton. Non-decisions are decisions made outside Easton's definition of the political, and this narrow viewpoint leads to the odd-sounding term. This is a good example of the power of metaphysics impinging upon empirical work.

Chapter 4: Political Power and Bargaining Theory

1   See Morris (1987) for a strong argument in favour of the earlier Penrose index and Riker (1969a, 1969b) for comparisons between some of the indices.

2   He also writes (1980, 349): 'power is the weight (metaphorically speaking) that an actor brings to the side he supports. To estimate an actor's power we need to find out how much of a disparity between the weight of the supports and the weight of the opponents he can reverse by taking part in the decision on a measure. ... To put it non-metaphorically, we can describe an actor's power by cataloguing the kinds of situation in which he can change a prospective loss to the side he supports to a win for it.'

3   On a frequency interpretation of probability it is quite standard, but my point is that the oddness of defining power as a probability should not really count against it too much.

4   Though Nagel's 'outcomes' only include affecting others. I allow outcome to be any possible social outcome ignoring resistance.

5   I prefer the term 'counteractual' to the more normal 'counterfactual' because it is confusing to call something which is a fact a 'counterfact'. If I kick you hard in the groin it will hurt. And that's a fact.

6   That is, according to the quotation in the text. The quotation in note 2 above indicates that individuals have power by the 'weight' they can bring to any situation, but even here it would be hard to gauge that weight across all possible worlds.

7   This is not to say that they are not costs, nor that they are not costs which will enter into A's calculation, just that they will not enter into A's calculation of the extent of her own power resulting from her legitimacy.

8   This statement needs to be qualified by the observation that if there are *general* cognitive biases in the way most people process information most of the time then these biases may need to be taken into account even in the general model. Many psychologists now believe that general biases do exist. See, for example, articles in Dowie and Lefrevre (eds), 1980.

Chapter 5: Collective Action and Dimensions of Power

1   It is not really true to say that Lukes 'created' the debate, but he is one of the most important and most cited exponents of the critique of behaviouralism.

Chapter 6: State Power Structures

1. There is a slight difficulty with this account. The latter clause suggests that changing policies is part of the functional logic, which must go beyond merely warming chairs. However, Jordan and Richardson say that 'it *should* produce more acceptable policies', not that it *does*. This suggests that it is the expectations engendered by the consultative process which are important rather than the actual result of those consultations. Of course, the functional argument only works whilst the system is maintained. Perhaps if expectations are never satisfied then the system will break down. This suggests that there must be some statistical relationship between group outcome power and system maintenance. However, this apparently empirical point could end up in the circularity which haunts all functional arguments.

2. Though a determinist thesis which explained outcomes purely by structural features of the state might be a thesis about state automation: the 'automatic state thesis'.

3. This statement confused a couple of readers since they felt that an oxymoron *is* a nonsensical statement. However, the *Concise Oxford English Dictionary* suggests otherwise: '**oxymoron** *n.* (Rhet.) Figure of speech with pointed conjunction of seemingly contradictory expressions.'

Chapter 7: Preference Formation, Social Location and Ideology

1. Justice is given in Boudon (1989). Although my views do not simply echo Boudon's I do not believe I contradict him on any important point.

2. Roemer just means the identification of P_t by the term 'equilibrium'.

3. The following story is a simplified version of that told in Smith, 1990a, 1990b.

4. Rescher, 1978, identifies five distinct types of 'existence', of which his second – the of-them correlative existence of ground-level universals such as the properties of particulars, be they actual or merely possible – applies to beliefs.

Part II: Postscript

1. I would like to thank Will Bosworth and Anne Gelling for comments on this Postscript. Will especially corrected various errors in the first draft, sometimes reminding me of what my own position is.

Chapter 9: Some Further Thoughts on Power

1. Later it attracted a lot more attention, much of it directed at the idea of luck and systematic luck (for example, Barry, 2002, 2003; Guzzini, 2005; Lukes and Haglund, 2005; Allen, 2008; Hindmoor and McGeechan, 2013).

2. Will Bosworth pointed out to me that Mary Wollstonecraft was the first person to make that claim.

3. Bittman et al. (2003) have data on employment levels and education, but do not attempt to match couples across these variables to find any such matching variation, looking only at relative earnings and domestic work at aggregate levels.

4. I somewhat confused the issue in later work, where I sometimes referred to these five ways in which power operates as resources. This was a mistake: they are ways in which power operates, whilst agential resources give agents the means by which to operate power in these ways.

5   Some 20 years ago, I embarked on just such an empirical study, which unfortunately never saw the light of day. With four other researchers and an ESRC Research grant, we were contracted to write a 120,000-word 'community power' study of London, covering the governance of London from the break-up of the GLC in 1986 to the election of the first Mayor in 2000. I wrote much of the first draft, which garnered enthusiastic reviewer comments. Two of my colleagues failed to deliver the chapters for which they were responsible, dealing a fatal blow to the project, though it did achieve some outputs (Dowding, 1996b, 2001; Dowding and Dunleavy, 1996; Dowding and King, 2000; Dowding et al., 1995; Dowding et al., 1999, 2000; Rydin, 1998a, 1998b).

Chapter 10: The Nature of the Exercise

1   I am aware of the personal irony here. My first published article, written as an undergraduate with my teacher Richard Kimber, is of just this ilk (Dowding and Kimber 1983). In my defence, it was not simply designed to define a term: our ulterior motive was to critique claims about political stability that transformed operationalizations of events that threaten stability with the concept of stability itself.

2   That is not the argument of Walter Gallie (1956), who coined the phrase 'essential contestability'; he thinks we also have to trace our disagreement back to an original exemplar, our disagreement being over the correct extension of that originally agreed term. See Evine (2014) for a good discussion of Gallie on this aspect.

3   In fact, being clones does not make them genetically identical since developmental factors also determine the precise genome of any token animal, but we will ignore that.

Bibliography

Agger, R.E., Goldrich, D. and Swanson, B.E. (1964), *The Rulers and the Ruled: Political Power and Impotence in American Communities*, New York, NY: Oxford University Press.

Allen, A. (1998), 'Rethinking Power', *Hypatia*, **13**, 21–40.

Allen, A. (1999), *The Power of Feminist Theory: Domination, Resistance, Solidarity*, Boulder, CO: Westview Press.

Allen, A. (2008), 'Rationalizing Oppression', *Journal of Power*, **1**, 1, 51–65.

Allison, G. (1971), *The Essence of Decision: Explaining the Cuban Missile Crisis*, Boston, MA: Little, Brown.

Arrow, K. (1963), *Social Choice and Individual Values*, 2nd edn, New York, NY: Wiley.

Atkinson, M.M. and Coleman, W.D. (1989), 'Strong States and Weak States: Sectoral Policy Networks in Advanced Capitalist Economies', *British Journal of Political Science*, **19**, 47–67.

Axelrod, R. (1970), *Conflict of Interest: A Theory of Divergent Goals with Applications to Politics*, Chicago, IL: Markham.

Bachrach, P. (1967), *The Theory of Democratic Elitism: A Critique*, Boston, MA: Little, Brown.

Bachrach, P. and Baratz, M. (1970), *Power and Poverty: Theory and Practice*, New York, NY: Oxford University Press.

Baldwin, D.A. (1989), *Paradoxes of Power*, Oxford: Basil Blackwell.

Barry, B. (1965), *Political Argument*, London: Routledge and Kegan Paul.

Barry, B. (1978), *Sociologists, Economists and Democracy*, 2nd edn, Chicago, IL: Chicago University Press.

Barry, B. (1980), 'Is It Better To Be Powerful or Lucky?', Parts 1 and 2, *Political Studies*, **28**, 183–94, 338–52.

Barry, B. (1988), 'Review Article: The Uses of "Power"', *Government and Opposition*, **23**, 340–53.

Barry, B. (1990), *Political Argument: A Reissue with a New Introduction*, New York, NY: Harvester Wheatsheaf.

Barry, B. (2002), 'Capitalists Rule OK? Some Puzzles about Power', *Politics, Philosophy and Economics*, **1**, 155–84.

Barry, B. (2003), 'Capitalists Rule. OK? A Commentary on Keith Dowding', *Politics, Philosophy and Economics*, **2**, 323–41.

Barry, B. and Hardin, R. (eds) (1982), *Rational Man and Irrational Society?*, Beverly Hills, CA: Sage.

Barry, B. and Rae, D. (1975), 'Political Evaluation', in Greenstein, D. and Polsby, N. (eds), *Political Science: Scope and Theory: Handbook of Political Science*, vol. 1, Reading, MA: Addison-Wesley.

Bartlett, R. (1989), *Economics and Power*, Cambridge: Cambridge University Press.

Bassett, K. and Harloe, M. (1990), 'Swindon: The Rise and Decline of a Growth Coalition', in Harloe, Pickvance and Urry (eds) (1990).

Bates, R.H., Greif, A., Levi, M., Rosenthal, J.-L. and Weingast, B.R. (1998), *Analytic Narratives*, Princeton, NJ: Princeton University Press.

Bates, R.H., Greif, A., Levi, M., Rosenthal, J.-L. and Weingast, B.R. (2000), 'The Analytic Narrative Project', *American Political Science Review*, **94**, 696–702.

Bealey, F., Blondel, J. and McCann, W. (1965), *Constituency Politics: A Study of Newcastle-Under-Lyme*, London: Faber.

Bechara, A., Damasio, H. and Damasio, A. (2000), 'Emotion, Decision Making and the Orbitofrontal Cortex', *Cerebral Coretex*, **10**, 295–307.

Becker, G. (1981) *A Treatise on the Family*, Cambridge, MA: Harvard University Press.

Bell, C. and Newby, H. (1971), *Community Studies*, London: Allen and Unwin.

Bell, R., Edwards, D.V. and Wagner, R.H. (eds) (1969), *Political Power: A Reader*, London: Collier-Macmillan.

Benewick, R., Berki, R.N. and Parekh, B. (1973), *Knowledge and Belief in Politics*, London: Allen and Unwin.

Benson, J.K. (1982), 'A Framework for Policy Analysis', in Rogers, D. and Whitten D. and Associates (eds), *Interorganizational Coordination*, Ames, IA: Iowa State University Press.

Bentley, A. (1967), *The Process of Government*, ed. Odegard, P. Cambridge, MA: Belknap Press (first published 1908).

Berlin, I. (1969), *Four Essays on Liberty*, Oxford: Oxford University Press.

Bhaskar, R. (1979), *Philosophy and the Human Sciences 1: The Possibility of Naturalism*, Brighton: Harvester.

Birch, A.H. (1984), 'Overload, Ungovernability and Delegitimation: The Theories and the British Case', *British Journal of Political Science*, **14**, 135–60.

Bittman, M., England, P., Sayer, L., Folbre, N. and Matheson, G. (2003), 'When Does Gender Trump Money? Bargaining and Time in Household Work', *American Journal of Sociology*, **109**, 1, 186–214.

Blalock, H. (1989), *Power and Conflict: Toward a General Theory*, Newbury Park, CA: Sage.

Blank, S. (1978), 'Britain: The Politics of Foreign Economic Policy, the Domestic Economy, and the Problem of Pluralistic Stagnation', in Katzenstein, P. (ed.), *Between Power and Plenty*, Madison, WI: University of Wisconsin Press.

Bosworth, W. (2016), 'An Interpretation of Political Argument', *European Journal of Political Theory*, online first, 1–21. doi: 10.1177/1474885116659842

Bosworth, W. and Dowding, K. (2019), 'The Cambridge School and Kripke: Bug Detecting with the History of Political Thought', *Review of Politics*, forthcoming (Sept).

Boudon, R. (1989), *The Analysis of Ideology*, Cambridge: Polity.

Bourdieu, P. (1991), *Language and Symbolic Power*, trans. by G. Raymond and M. Adamson, Cambridge, MA: Harvard University Press.

Bourdieu, P. (2001), *Masculine Domination*, trans. by Richard Nice, Stanford, CA: Stanford University Press.

Bridgman, P.W. (1927), *The Logic of Modern Physics*, New York, NY: W.W. Norton.

Brittan, S. (1975), 'The Economic Contradictions of Democracy', *British Journal of Political Science*, **5**, 129–59.

Bruner, J. (2015), 'Diversity, Tolerance and the Social Contract', *Politics, Philosophy and Economics*, **14**, 4, 429–48.

Buchanan, J.M. (1977), *Freedom in Constitutional Contract*, College Station, TX: Texas A. and M. Press.

Carling, A. (1991), *Social Division*, London: Verso.

Caro, R. (1974), *The Power Broker*, New York, NY: Alfred A. Knopf.

Chalmers, D.J. (2011), 'Verbal Disputes', *Philosophical Review*, **120**, 515–66.

Chalmers, D.J. (2012), *Constructing the World*, Oxford: Oxford University Press.

Chong, D. and Druckman, J.N. (2010), 'Dynamic Public Opinions: Communication Effects over Time', *American Political Science Review*, **104**, 663–80.

Coates, D. (1984), *The Context of British Politics*, London: Hutchinson.

Cohen, G.A. (1978), *Karl Marx's Theory of History: A Defence*, Oxford: Clarendon Press.

Cohen, G.A. (1989), 'On the Currency of Egalitarian Justice', *Ethics*, **99**, 906–44.

Collard, D. (1981), *Altruism and Economy: A Study in Non-Selfish Economics*, Oxford: Martin Robertson.

Connolly, W.E. (1983), *The Terms of Political Discourse*, 2nd edn, Oxford: Martin Robertson.

Crenson, M. (1971), *The Unpolitics of Air Pollution*, Baltimore, MD: Johns Hopkins University Press.

Cudd, A. (2005), 'How to Explain Oppression: Criteria for Adequacy for Normative Explanatory Theories', *Philosophy of Social Sciences*, **35**, 1, 20–49.

Cudd, A. (2006), *Analyzing Oppression*, Oxford: Oxford University Press.

Dahl, R.A. (1956), *A Preface to Democratic Theory*, Chicago, IL: Chicago University Press.

Dahl, R.A. (1957), 'The Concept of Power', *Behavioral Science*, **2**, 201–15.

Dahl, R.A. (1961a), *Who Governs? Democracy and Power in an American City*, New Haven, CT: Yale University Press.

Dahl, R.A. (1961b), 'The Behavioral Approach in Political Science: Epitaph for a Monument to a Successful Protest', *American Political Science Review*, **55**, 763–72.

Dahl, R.A. (1961c), 'Equality and Power in American Society', in D'Antonio, W.V. and Ehlrich, H.J. (eds), *Power and Democracy in America*, South Bend, IN: Notre Dame University Press.

Dahl, R.A. (1963), 'Reply to Thomas Anton's "Power, Pluralism and Local Politics"', *Administrative Science Quarterly*, **7**, 250–6.

Dahl, R.A. (1968), 'Power', in Sils, D.L. (ed.), *International Encylopaedia of the Social Sciences*, vol. 12, New York, NY: Free Press.

Dahl, R.A. (1969a), 'A Critique of the Ruling Elite Model', in Bell et al. (eds) (1969), 36–41 (first published 1958).

Dahl, R.A. (1969b), 'The Concept of Power', in Bell et al. (eds) (1969), pp. 79–93 (first published 1957).

Dahl, R.A. (1986), 'Rethinking *Who Governs?*: New Haven Revisited', in Waste, R. (ed.) (1986).

Damasio, A. (1994), *Descartes' Error: Emotion, Reason and the Human Brain*, New York, NY: Putnam.

Davidson, D. (1980), *Essays on Actions and Events*, Oxford: Clarendon Press.

Davidson, D. (1982), 'Paradoxes of Irrationality', in Wollheim, R. and Hopkins, J. (eds), *Philosophical Essays on Freud*, Cambridge: Cambridge University Press.

Davidson, D. (1985), 'A New Basis for Decision Theory', *Theory and Decision*, **18**, 87–98.

Day, J.P. (1987), 'Threats, Offers, Law, Opinion and Liberty', in his *Liberty and Justice*, London: Croom Helm.

Dearlove, J. (1973), *The Politics of Policy in Local Government*, London: Cambridge University Press.

Debnam, G. (1975), 'Nondecisions and Power: The Two Faces of Bachrach and Baratz', *American Political Science Review*, **69**, 889–99.

Debnam, G. (1985), *The Analysis of Power: A Realist Approach*, London: Macmillan.

Dennett, D.C. (1998), 'Real Patterns', in his *Brainchildren: Essays on Designing Minds*, Harmondsworth: Penguin.

Deutsch, K. (1966), *The Nerves of Government*, 2nd edn, New York, NY: Free Press.

Domhoff, G.W. (1978), *Who Really Rules? New Haven and Community Power Reexamined*, Brunswick, NJ: Transaction.

Domhoff, G.W. (1983), *Who Rules America Now?*, Englewood Cliffs, NJ: Prentice-Hall.

Domhoff, G.W. (1986), 'The Growth Machine and the Power Elite: A Challenge to Pluralists and Marxists Alike', in Waste, R. (ed.) (1986).

Domhoff, G.W. and Dye, T.R. (eds) (1987), *Power Elites and Organizations*, Beverly Hills, CA: Sage.

Dowding, K. (1987), 'Collective Action, Group Organization and Pluralist Democracy', DPhil: University of Oxford.

Dowding, K. (1990), 'Ability and Ableness: Morriss on Power and Counteractuals', *Government Department Working Papers*, **10**, Uxbridge: Brunel University.

Dowding, K. (1996a), *Power*, Buckingham and Minneapolis, MN: Open University Press and Minnesota University Press.

Dowding, K. (1996b), 'Public Choice and Local Governance', in King, D. and Stoker, G. (eds), *Rethinking Local Democracy*, London: Macmillan.

Dowding, K. (1999), 'Shaping Future Luck', *Journal of Conflict Processes and Change*, **4**, 1–11, reprinted in Dowding (2017), ch. 6.

Dowding, K. (2001), 'Explaining Urban Regimes', *International Journal of Urban and Regional Research*, **25**, 7–19.

Dowding, K. (2003), 'Resources, Power and Systematic Luck: Reply to Barry', *Politics, Philosophy and Economics*, **3**, 305–22, reprinted in Dowding (2017), ch. 5.

Dowding, K. (2006), 'Three Dimensional Power: A Discussion of Steven Lukes Power: A Radical View Second Edition', *Political Studies Review*, **4**, 136–45.

Dowding, K. (2008), 'Agency and Structure: Interpreting Power Relationships', *Journal of Power*, **1**, 21–36, reprinted in Dowding (2017), ch 2.

Dowding, K. (2010), 'Luck, Equality and Responsibility', *Critical Review of International Social and Political Philosophy*, **13**, 71–92, reprinted in Dowding (2017), ch 7.

Dowding, K. (2012), 'Why Should We Care About the Definition of Power?', *Journal of Political Power*, **5**, 119–35, reprinted in Dowding (2017), ch. 1.

Dowding, K. (2013), 'The Role of Political Argument in Justice as Impartiality', *Political Studies*, **61**, 67–81.

Dowding, K. (2016a), *The Philosophy and Methods of Political Science*, London: Palgrave.

Dowding, K. (2016b), 'Power and Persuasion', *Political Studies*, **64**, 1–15.

Dowding, K. (2017), *Power, Luck and Freedom: Collected Essays*, Manchester: Manchester University Press.

Dowding, K. (2018), 'Emotional Appeals in Politics and Deliberation', *Critical Review of International Social and Political Philosophy*, **21**, 2, 242–60.

Dowding, K. (forthcoming), 'Can a Case-Study Test a Theory? Types and Tokens in Social Explanation', in Peters, B.G. and Fontaine, G. (eds), *Handbook of Methods for Comparative Policy Analysis*, Aldershot: Edward Elgar.

Dowding, K. and Bosworth, W. (2018), 'Ambiguity and Vagueness in Political Terminology: On Coding and Referential Vacuity', *European Journal of Political Theory*, online first. doi: 10.1177/1474885118771256

Dowding, K. and Dunleavy, P. (1996), 'Production, Disbursement and Consumption: The Modes and Modalities of Goods and Services', in Edgell, S., Hetherington, K. and Warde A. (eds), *Consumption Matters*, Oxford: Blackwell.

Dowding, K. and Hindmoor, A. (1997), 'The Usual Suspects: Rational Choice, Socialism and Political Theory', *New Political Economy*, **2**, 451–63.

Dowding, K. and Kimber, R. (1983), 'The Meaning and Use of "Political Stability"', *European Journal of Political Research*, **11**, 229–43.

Dowding, K. and Kimber, R. (1987), 'Political Stability and the Science of Comparative Politics', *European Journal for Political Research*, **15**, 103–22.

Dowding, K. and King, D. (2000), 'Rooflessness in London', *Policy Studies Journal*, **28**, 365–81.

Dowding, K. and Miller, C. (forthcoming), 'On Prediction in Political Science', *European Journal of Political Research*. doi: 10.1111/1475-6765 .12319

Dowding, K. and van Hees, M. (2009), 'Freedom of Choice', in Anand, P., Pattanaik, P.K. and Puppe, C. (eds), *Oxford Handbook of Rational and Social Choice*, 374–92, Oxford: Oxford University Press.

Dowding, K., Dunleavy, P., King, D. and Margetts, H. (1995), 'Rational Choice and Community Power Structures', *Political Studies*, **43**, 265–77, reprinted in Dowding (2017), ch. 3.

Dowding, K., Dunleavy, P., King, D., Margetts, H. and Rydin, Y. (1999), 'Regime Politics in London Local Government', *Urban Affairs Review*, **34**, 515–45.

Dowding, K., Dunleavy, P., King, D., Margetts, H. and Rydin, Y. (2000), 'Understanding Urban Governance: The Contribution of Rational Choice', in Stoker, G. (ed), *The New Politics of British Local Governance*, 91–116, Houndmills: Macmillan.

Dowie, J. and Lefrevre, J. (eds) (1980), *Risk and Chance: Selected Readings*, Milton Keynes: Open University Press.

Downs, A. (1957), *An Economic Theory of Democracy*, New York, NY: Harper and Row.

Druckman, J.N., Fein, J. and Leeper, T.J. (2012), 'A Source of Bias in Public Opinion Stability', *American Political Science Review*, **106**, 430–54.

Dummett, M. (1984), *Voting Procedures*, Oxford: Clarendon Press.

Dunleavy, P. (1981a), *The Politics of Mass Housing in Britain*, Oxford: Clarendon Press.

Dunleavy, P. (1981b), 'Professions and Policy Change: Notes Towards a Model of Ideological Corporatism', *Public Administration Bulletin*, **36**, 3–16.

Dunleavy, P. (1985), 'Bureaucrats, Budgets and the Growth of the State: Reconstructing an Instrumental Model', *British Journal of Political Science*, **15**, 299–328.

Dunleavy, P. (1986), 'Explaining the Privatization Boom: Public Choice versus Radical Approaches', *Public Administration*, **64**, 13–34.

Dunleavy, P. (1988), 'Group Identities and Individual Influence: Re-constructing the Theory of Interest Groups', *British Journal of Political Science*, **18**, 21–50.

Dunleavy, P. (1989), 'The Architecture of the British Central State', *Public Administration*, **67**, 'Part I: Framework for Analysis', 249–75, 'Part II: Empirical Findings', 391–417.

Dunleavy, P. and O'Leary, B. (1987), *Theories of the State*, Basingstoke: Macmillan Educational Ltd.

Dunleavy, P. and Ward, H. (1981), 'Exogenous Voter Preferences and Parties with State Power: Some Internal Problems with Economic Theories of Party Competition', *British Journal of Political Science*, **11**, 350–80.

Dye, T.R. (1986), 'Community Power and Public Policy', in Waste, R. (ed.) (1986).

Easton, D. (1953), *The Political System*, New York, NY: Knopf.

Easton, D. (1965), *A Systems Analysis of Political Life*, New York, NY: Wiley.

Easton, D. (1966), *A Framework for Political Analysis*, Chicago, IL: Chicago University Press.

Elster, J. (1978), *Logic and Society*, London: Wiley.

Elster, J. (1983a), *Sour Grapes: Studies on the Subversion of Rationality*, Cambridge: Cambridge University Press.

Elster, J. (1983b), *Explaining Technical Change*, Cambridge: Cambridge University Press.

Elster, J. (1984), *Ulysses and the Sirens: Studies in Rationality and Irrationality*, 2nd edn, Cambridge: Cambridge University Press.

Elster, J. (1985), *Making Sense of Marx*, Cambridge: Cambridge University Press.

Elster, J. (1986), 'Introduction' to Elster, J. (ed.), *Rational Choice*, Oxford: Basil Blackwell, 1–33.

Elster, J. (1988), 'Marx, Revolution and Rational Choice', in Taylor, M. (ed.), *Rationality and Revolution*, Cambridge: Cambridge University Press.

Evnine, S.J. (2014), 'Essentially Contested Concepts and Semantic Externalism', *Journal of the Philosophy of History*, **8**, 118–40.

Finer, S.E. (1966), *Anonymous Empire*, 2nd edn, London: Pall Mall.

Follett, M.P. (1942), 'Power', in Metcalf, H.C. and Urwick, L. (eds), *Dynamic Administration: The Collected Papers of Mary Parker Follett*, New York, NY: Harper.

Foucault, M. (2015), *The History of Sexuality: Vol 1: The Will to Knowledge*, London: Penguin.

Frey, F. (1971), 'Comment: On Issues and Nonissues in the Study of Power', *American Political Science Review*, **65**, 1081–1101.

Friedland, R. (1982), *Power and Crisis in the City*, London: Macmillan.

Friedrich, C. (1941), *Constitutional Government and Democracy*, New York, NY: W. W. Norton.

Frohlich, N. (1974), 'Self-Interest or Altruism: What Difference?', *Journal of Conflict Resolution*, **18**, 55–73.

Gallie, W.B. (1956), 'Essentially Contested Concepts', *Proceedings of the Aristotelian Society*, **56**, 167–98.

Gaventa, J. (1980), *Power and Powerlessness: Quiescence and Rebellion in an Appalachian Valley*, Oxford: Clarendon Press.

Gerring, J. (1999), 'What Makes a Concept Good? A Criterial Framework for Understanding Concept Formation in the Social Sciences', *Polity*, **31**, 3, 357–93.

Giddens, A. (1984), *The Constitution of Society*, Oxford: Polity Press.

Goodin, R. (1980), *Manipulatory Politics*, New Haven, CT: Yale University Press.

Goodin, R. and Dryzek, J. (1981), 'Rational Participation: The Politics of Relative Power', *British Journal of Political Science*, **10**, 273–92.

Grafstein, R. (1988), 'A Realist Foundation for Essentially Contested Concepts', *Western Political Quarterly*, **41**, 9–28.

Graham, A. (1972), 'Industrial Policy', in Beckerman, W. (ed.), *The Government's Economic Record 1964–70*, London: Duckworth.

Grandy, R. (1973), 'Reference, Meaning and Belief', *Journal of Philosophy*, LXX, 439–52.

Grant, W. (1978), 'Insider Groups, Outsider Groups and Interest Group Strategies in Britain', *Department of Politics, Working Papers*, **18**, Coventry: University of Warwick.

Grant, W. (1987), *Business and Politics in Britain*, London: Macmillan.

Gray, J. (1977), 'On the Contestability of Social and Political Concepts', *Political Theory*, **5**, 331–48.

Gray, J. (1979), 'On Liberty. Liberalism and Essential Contestability', *British Journal of Political Science*, **8**, 385–402.

Grofman, B. and Scarrow, H. (1979), 'Ianucci and its Aftermath: The Application of the Banzhaf Index to Weighted Voting in the State of New York', in Barrow, S.J., Schotter, A. and Schwodiater, G. (eds), *Applied Game Theory*, Wurzburg: Physica Verlag.

Guzzini, S. (2005), 'The Concept of Power: A Constructivist Analysis', *Millennium: Journal of International Studies*, **33**, 495–521.

Hamilton, M.G. (1987), 'The Elements of the Concept of Ideology', *Political Studies*, **35**, 18–38.

Harding, A. (1990), 'Property Interests and Urban Growth Coalitions in the U.K.: A Brief Encounter?', *CUS Working Paper*, **12**, University of Liverpool.

Harloe, M., Pickvance, C. and Urry, J. (eds) (1990), *Place, Policy and Politics: Do Localities Matter?*, London: Unwin Hyman.

Harsanyi, J.C. (1962a), 'Measurement of Social Power in N-Person Recriprocal Power Situations', *Behavioral Science*, **7**, 81–91.

Harsanyi, J.C. (1962b), 'Measurement of Social Power, Opportunity Costs, and the Theory of Two-Person Bargaining Games', *Behavioral Science*, **7**, 67–80.

Harsanyi, J.C. (1969a), 'Measurement of Social Power, Opportunity Costs, and the Theory of Two-Person Bargaining Games', in Bell et al. (eds) (1969), 226–38 and in Harsanyi (1976).

Harsanyi, J.C. (1969b), 'Measurement of Social Power in n-Person Reciprocal Power Situations', in Bell et al. (eds) (1969), 239–48 and in Harsanyi (1976).

Harsanyi, J.C. (1976), *Essays on Ethics, Social Behavior and Scientific Explanation*, Dordrecht: D. Reidel.

Harsanyi, J.C. (1977), *Rational Behavior and Equilibrium in Games and Social Situations*, Cambridge: Cambridge University Press.

Heath, J. (2000), 'Ideology, Irrationality and Collectively Self-Defeating Behavior', *Constellations*, **7**, 3, 363–71.

Heclo, H. (1978), 'Issue Networks and the Executive Establishment', in King, A. (ed.), *The New American Political System*, Washington, DC: American Enterprise Institute.

Hedstrom, P. and Ylikoski, P. (2010), 'Causal Mechanisms in the Social Sciences', *Annual Review of Sociology*, **36**, 49–67.

Held, D. (1987), *Models of Democracy*, Oxford: Polity Press.

Henderson, P.D. (1977), 'Two British Errors: Their Probable Size and Some Possible Lessons', *Oxford Economic Papers*, **29**, 159–205.

Hindess, B. (1982), 'Power, Interests and the Outcomes of Struggles', *Sociology*, **16**, 498–511, reprinted in Hindess (1989).

Hindess, B. (1988), *Choice, Rationality and Social Theory*, London: Unwin Hyman.

Hindess, B. (1989), *Political Choice and Social Structure*, Aldershot: Edward Elgar.

Hindmoor, A. and McGeechan, J. (2013), 'Luck, Systematic Luck and Business Power: Lucky All the Way Down or Trying Hard to Get What It Wants without Trying', *Political Studies*, **61**, 834–50.

Hirschman, A.O. (1970), *Exit, Voice and Loyalty: Responses to Decline in Firms, Organizations and States*, Cambridge, MA: Harvard University Press.

Hobsbawm, E.J. (1952), 'The Machine Breakers', *Past and Present*, **1**, 57–70.

Hollis, M. and Nell, E.J. (1975), *Rational Economic Man: A Philosophical Critique of Neo-Classical Economics*, Cambridge: Cambridge University Press.

Hudson, R. (1990), 'Trying to Revive an Infant Hercules: The Rise and Fall of Local Authority Modernization Policies on Teesside', in Harloe, Pickvance and Urry (eds) (1990).

Hunter, F. (1953), *Community Power Structure*, Chapel Hill, NC: University of North Carolina Press.

Ibsen, H. (1960), *An Enemy of the People*, trans. by J.W. McFarlane, London: Oxford University Press.

Ingham, G. (1984), *Capitalism Divided? The City and Industry in British Social Development*, London: Macmillan.

Isaac, J. (1987), *Power and Marxist Theory: A Realist View*, Ithaca, NY: Cornell University Press.

Jennings, M. (1964), *Community Influentials: The Elites of Atlanta*, New York, NY: Free Press.

Jones, G.W. (1969), *Borough Politics: A Study of Wolverhampton Town Council, 1888–1964*, London: Macmillan.

Jordan, G. (1990a), 'Sub-governments, Policy Communities and Networks: Refilling the Old Bottles', *Journal of Theoretical Politics*, **2**, 319–38.

Jordan, G. (1990b), 'Policy Community Realism versus "New" Institutionalist Ambiguity', *Political Studies*, **38**, 470–84.

Jordan, G. and Richardson, J.J. (1982), 'The British Policy Style or the Logic of Negotiation?', in Richardson, J.J. (ed.), *Policy Styles in Western Europe*, Hemel Hempstead: George Allen and Unwin.

Jordan, G. and Richardson, J.J. (1987a), *British Politics and the Policy Process*, London: Unwin Hyman.

Jordan, G. and Richardson, J.J. (1987b), *Government and Pressure Groups in Britain*, Oxford: Oxford University Press.

Kay, P. and Kempton, W. (1984), 'What Is the Sapir-Whorf-Hypothesis?', *American Anthropologist*, **86**, 65–79.

Kimber, R. and Richardson, J.J. (1974), 'The Roskillers: Cublington Fights the Airport', in Kimber and Richardson (eds) (1974).

Kimber, R. and Richardson, J.J. (eds) (1974), *Campaigning for the Environment*, London: Routledge and Kegan Paul.

King, D.S. (1990), 'Economic Activity and the Challenge to Local Government', in King, D.S. and Pierre, J. (eds), *Challenges to Local Government*, London: Sage.

King, D.S. and Wickham-Jones, M. (1990), 'Social Democracy and Rational Workers', *British Journal of Political Science*, **20**, 387–413.

Kogan, M. (1975), *Educational Policy-Making*, London: Allen and Unwin.

Krantz, D.H., Duncan Luce, R., Suppes, P. and Tversky, A. (1971), *Foundations of Measurement, Vol 1*, New York: Academic Press.

Kreps, D.M., Milgram, P., Roberts, J. and Wilson, W. (1982), 'Rational Cooperation in the Finitely Repeated Prisoner's Dilemma', *Journal of Economic Theory*, **27**, 245–52.

LaCroix, T. and O'Connor, C. (2018), 'Power by Association', *MS*, http://philsci-archive.pitt.edu/14318.

Lange, P. (1984), 'Unions, Workers, and Wage Regulations: the Rational Bases of Consent', in Goldthorpe, J. (ed.), *Order and Conflict in Contemporary Capitalism*, Oxford: Oxford University Press.

Laver, M. and Schofield, N. (1990), *Multiparty Government: The Politics of Coalition Government in Europe*, Oxford: Oxford University Press.

Levine, A., Sober, E. and Wright, E.O. (1987), 'Marxism and Methodological Individualism', *New Left Review*, **162**, 64–87.

Lewis, D. (1970), 'How to Define Theoretical Terms', *Journal of Philosophy*, **67**, 13, 427–46.

Lindblom, C.E. (1977), *Politics and Markets*, New York, NY: Basic Books.

List, C. and Spiekermann, K. (2013), 'Methodological Individualism and Holism in Political Science: A Reconciliation', *American Political Science Review*, **107**, 629–43.

Logan, J. and Molotch, H. (1984), 'Tensions in the Growth Machine: Overcoming Resistance to Value-Free Development', *Social Problems*, **31**, 483–99.

Logan, J. and Molotch, H. (1987), *Urban Fortunes: The Political Economy of Place*, Berkeley, CA: University of California Press.

Lowe, P. and Goyder, J. (1983), *Environmental Groups in Politics*, London: Allen and Unwin.

Lukes, S. (1974), *Power: A Radical View*, London: Macmillan.

Lukes, S. (1977), *Essays in Social Theory*, London: Macmillan.

Lukes, S. (1991), *Moral Conflict and Politics*, Oxford: Clarendon Press.

Lukes, S. (2005), *Power: A Radical View: The Original Text with Two Major New Chapters*, Basingstoke: Palgrave Macmillan.

Lukes, S. and Haglund, L. (2005), 'Power and Luck', *Archives Européennes de Sociologie*, **46**, 45–66.

McElroy, M.B. and Horney, M.J. (1981), 'Nash Bargained Household Decisions: A Generalization of the Theory of Demand', *International Economic Review*, **22**, 2, 333–49.

MacIntyre, A. (1973), 'The Essential Contestability of Some Social Concepts', *Ethics*, **84**, 1–19.

Mackay, A.F. (1980), *Arrow's Theorem: The Paradox of Social Choice*, New Haven, CT: Yale University Press.

Mackie, G. (1996), 'Ending Footbinding and Infibulation: A Convention Account', *American Sociological Review*, **61**, 999–1017.

Mackie, J. (1974), *The Cement of the Universe*, Oxford: Oxford University Press.

McLean, I. (1987), *Introduction to Public Choice*, Oxford: Basil Blackwell.

Macridis, R. (1961), 'Interest Groups in Comparative Analysis', *Journal of Politics*, **23**, 25–45.

Mahony, R. (1995), *Kidding Ourselves: Breadwinning, Babies, and Bargaining Power*, New York, NY: Basic Books.

Manser, M. and Brown, M. (1980), 'Marriage and Household Decision-Making: A Bargaining Analysis', *International Economic Review*, **21**, 1, 31–44.

Margolis, H. (1982), *Selfishness, Altruism and Rationality*, Cambridge: Cambridge University Press.

Marsh, D. (1983), 'Interest Group Activity and Structural Power: Lindblom's *Politics and Markets*', *West European Politics*, **6**, 3–13.

Marsh, D. and Locksley, G. (1983), 'Capital in Britain: Its Structural Power and Influence over Policy', *West European Politics*, **6**, 37–60.

May, T. and Nugent, N. (1982), 'Insiders, Outsiders and Thresholders', paper presented at Political Studies Association University of Kent, 14–16 April.

Merrelman, R.M. (1968), 'On the Neo-Elitist Critique of Community Power', *American Political Science Review*, **62**, 451–60.

Miller, D. (1976), *Social Justice*, Oxford: Oxford University Press.

Miller, D. (1983), 'Linguistic Philosophy and Political Theory', in Miller, D. and Siedentop, L. (eds), *The Nature of Political Theory*, Oxford: Oxford University Press.

Mills, C.W. (1956), *The Power Elite*, New York, NY: Oxford University Press.

Moe, T.M. (1980), *The Organization of Interests*, Chicago, IL: University of Chicago Press.

Molotch, H. (1976), 'The City as a Growth Machine', *American Journal of Sociology*, **82**, 309–30.

Molotch, H. (1979), 'Capital and Neighbourhood in the United States', *Urban Affairs Quarterly*, **14**, 289–312.

Moran, M. (1983), 'Power, Policy and the City of London', in King, R. (ed.), *Capital and Politics*, London: Routledge and Kegan Paul.

Morriss, P. (1972), 'Power in New Haven: A Reassessment of "Who Governs?"', *British Journal of Political Science*, **2**, 457–65.

Morriss, P. (1987), *Power: A Philosophical Analysis*, Manchester: Manchester University Press.

Morriss, P. (2002), *Power: A Philosophical Analysis*, 2nd edn, Manchester: Manchester University Press.

Nagel, J. (1975), *The Descriptive Analysis of Power*, New Haven, CT: Yale University Press.

Newton, K. (1976), *Second City Politics*, Oxford: Oxford University Press.

Newton, K. (1979), 'The Language and the Grammar of Political Power: A Comment on Polsby', *Political Studies*, **27**, 542–7.

Niskanen, W.A. (1971), *Bureaucracy and Representative Government*, Chicago, IL: Aldine.

Nordlinger, E.A. (1981), *On the Autonomy of the Democratic State*, Cambridge, MA: Harvard University Press.

Nozick, R. (1977), 'On Austrian Methodology', *Synthese*, **36**, 353–92.

O'Connor, C. (2019), *The Origins of Unfairness*, Oxford: Oxford University Press.

Offe, C. and Wiesenthal, H. (1985), 'Two Logics of Collective Action: Theoretical Notes on Social Class and Organizational Form' in Offe, C., *Disorganized Capitalism*, Oxford: Polity Press.

Okin, S.M. (1989), *Justice, Gender and the Family*, New York, NY: Basic Books.

Olson, M. (1971), *The Logic of Collective Action: Public Goods and the Theory of Groups*, Cambridge, MA: Harvard University Press.

Olson, M. (1982), *The Rise and Decline of Nations*, New Haven, CT: Yale University Press.

Oppenheim, F. (1981), *Political Concepts: A Reconstruction*, Oxford: Basil Blackwell.

Ostrom, E. (1990), *Governing the Commons: The Evolution of Institutions for Collective Action*, Cambridge: Cambridge University Press.

Ostrom, E. (2001), 'Decentralization and Development: The New Panacea', in Dowding, K., Hughes, J. and Margetts, H. (eds), *Challenges to Democracy*, Houndmills: Palgrave.

Ostrom, E. (2005), *Understanding Institutional Diversity*, Princeton, NJ: Princeton University Press.

Pansardi, P. (2011), 'Power To and Power Over', in Dowding, K. (ed), *Encyclopedia of Power*, 521–24, Beverly Hills, CA: Sage.

Parkhurst, G.M., Shogren, J.F. and Bastian, C. (2004), 'Repetition, Communication, and Coordination Failure', *Experimental Economics*, **7**, 2, 141–52.

Parry, G. and Morriss, P. (1974), 'When is a Decision Not a Decision?', in Crewe, I. (ed.), *British Sociology Yearbook Volume 1: Elites in Western Democracy*, London: Croom Helm.

Peterson, P.E. (1981), *City Limits*, Chicago, IL: University of Chicago Press.

Pettit, P. (1978), 'Rational Man Theory', in Hookway, C. and Pettit, P. (eds), *Action and Interpretation*, Cambridge: Cambridge University Press.

Pettit, P. (1997), *Republicanism: A Theory of Freedom and Government*, Oxford: Oxford University Press.

Pettit, P. (2012), *On the People's Terms: A Republican Theory and Model of Democracy*, Cambridge: Cambridge University Press.

Polsby, N. (1969), 'How to Study Community Power: The Pluralist Alternative', in Bell et al. (eds) (1969), 3–35 (first published 1960).

Polsby, N. (1979), 'Empirical Investigation of the Mobilization of Bias in Community Power Research', *Political Studies*, **27**, 527–41.

Polsby, N. (1980), *Community Power and Political Theory*, 2nd edn, New Haven, CT: Yale University Press.

Popper, K. (1972), *The Logic of Scientific Discovery*, 5th edn, London: Hutchinson.

Putnam, H. (1978), *Meaning and the Moral Sciences*, Boston, MA: Routledge and Kegan Paul.

Przeworski, A. (1986), *Capitalism and Social Democracy*, Cambridge: Cambridge University Press.

Przeworski, A. and Sprague, J. (1985), *Paper Stones: A History of Electoral Socialism*, Chicago, IL: University of Chicago Press.

Rescher, N. (1975), *A Theory of Possibility*, Oxford: Basil Blackwell.

Rescher, N. (1978), 'The Equivocality of Existence', in Rescher, N. (ed.), *Studies in Ontology: American Philosophical Quarterly Monograph Series*, **12**, Oxford: Basil Blackwell.

Rhodes, R.A.W. (1988), *Beyond Westminster and Whitehall*, London: Unwin Hyman.

Rhodes, R.A.W. (1990), 'Policy Networks: A British Perspective', *Journal of Theoretical Politics*, **2**, 293–317.

Ricci, D. (1971), *Community Power and Democratic Theory: The Logic of Political Analysis*, New York, NY: Random House.

Richardson, J.J. and Jordan, G. (1979), *Governing under Pressure: The Policy Process in a Post-Parliamentary Democracy*, Oxford: Martin Robertson.

Richardson, J.J. and Watts, N. (1985), *National Policy Styles and the Environment*, Berlin: Physica Verlag.

Riker, W. (1962), *The Theory of Political Coalitions*, New Haven, CT: Yale University Press.

Riker, W. (1969a), 'Some Ambiguities in the Notion of Power', in Bell et al. (eds) (1969), 110–19 (first published 1964).

Riker, W. (1969b), 'A Test of the Adequacy of the Power Index', in Bell et al. (eds) (1969), 214–25.

Riker, W.H. (1982), *Liberalism Against Populism: A Confrontation between the Theory of Democracy and the Theory of Social Choice*, San Francisco, CA: W.H. Freeman and Co.

Roberson, D., Davies, I. and Davidoff, J. (2000), 'Colour Categories Are Not Universal: Replications and New Evidence in Favour of Linguistic Relativity', *Journal of Experimental Psychology: General*, **129**, 369–98.

Roberts, R. (1985), 'Reputations in Games and Markets', in Roth, A.E. (ed.), *Game-Theoretic Models of Bargaining*, Cambridge: Cambridge University Press.

Roemer, J.E.(1986a), 'Equality of Resources Implies Equality of Welfare', *Quarterly Journal of Economics*, **101**, 751–84.

Roemer, J.E.(1986b), '"Rational Choice" Marxism: Some Issues of Method and Substance', in Roemer, J.E. (ed.), *Analytical Marxism*, 191–201, Cambridge: Cambridge University Press.

Roemer, J.E. (1998), *Equality of Opportunity*. Cambridge, MA: Harvard University Press.

Roemer, J.E. (2002), 'Equality of Opportunity: A Progress Report', *Social Choice and Welfare*, **19**, 455–71.

Rousseau, J.-J. (1984), *A Discourse on Inequality*, trans. and ed. Cranston, M., Harmondsworth: Penguin.

Rubinstein, A. (1982), 'Perfect Equilibrium in a Bargaining Model', *Econometrica*, **50**, 97–109.

Rubinstein, A. (1985), 'A Bargaining Model with Incomplete Information about Time Preferences', *Econometrica*, **53**, 1151–72.

Runciman, W. (1974), 'Relativism: Cognitive and Moral', *Proceedings of the Aristotelian Society: Supplementary Volume*, **68**, 191–208.

Ryan, M. (1978), *The Acceptable Pressure Group*, Farnborough: Saxon House.

Rydin, Y. (1998a), '"Managing Urban Air Quality": Language and Rational Choice in Metropolitan Governance', *Environment and Planning A: Economy and Space*, **30**, 8, 1429–43.

Rydin, Y. (1998b), 'The Enabling Local State and Urban Development: Resources, Rhetoric and Planning in East London', *Urban Studies*, **35**, 175–91.

Ryle, G. (1949), *The Concept of Mind*, Harmondsworth: Penguin.

Saunders, P. (1979), *Urban Politics: A Sociological Interpretation*, London: Hutchinson.

Schattschneider, E.E. (1960), *The Semi-Sovereign People: A Realist's View of Democracy in America*, New York, NY: Holt, Rinehart and Winston.

Schelling, T.C. (1966), *Arms and Influence*, New Haven, CT: Yale University Press.

Schelling, T.C. (1982), 'Hockey Helmets, Daylight Saving, and Other Binary Choices', in Barry and Hardin (eds) (1982).

Schick, F. (1982), 'Under What Descriptions?', in Sen, A. and Williams, B. (eds), *Utilitarianism and Beyond*, Cambridge: Cambridge University Press.

Schofield, N. (1985), 'Anarchy, Altruism and Cooperation: A Review', *Social Choice and Welfare*, **2**, 3, 207–19.

Segal, J.A. and Cover, A.D. (1989), 'Ideological Values and the Votes of U.S. Supreme Court Justices', *American Political Science Review*, **83**, 557–65.

Self, P. and Storing, J. (1962), *The State and the Farmer*, London: George Allen and Unwin.

Sen, A. (1982), *Choice, Welfare and Measurement*, Oxford: Basil Blackwell.

Sen, A. (1982a), 'Choice Functions and Revealed Preference', in Sen (1982).

Sen, A. (1982b), 'Behaviour and the Concept of Preference', in Sen (1982).

Sen, A. (1982c), 'Choice, Orderings and Morality', in Sen (1982).

Sen, A. (1982d), 'Rational Fools: A Critique of the Behavioural Foundations of Economic Theory', in Sen (1982).

Shapley, L.S. (1967), 'On Committees', in Zwicky, F. and Wilson, A.G. (eds), *New Methods of Thoughts and Procedure*, New York, NY: Springer-Verlag.

Shapley, L.S. (1981), 'Measurement of Power in Political Systems', *Proceedings of Symposia in Applied Mathematics*, **24**, 69–81.

Shapley, L.S. and Shubik, M. (1969), 'A Method for Evaluating the Distribution of Power in a Committee System', in Bell et al. (eds) (1969), 209–13.

Sharpe, J. and Newton, K. (1984), *Does Politics Matter?*, Oxford: Clarendon.

Simon, ILA. (1969), 'Notes on the Observation and Measurement of Power', in Bell et al. (eds) (1969), 69–78.

Skinner, B.F. (1953), *Science and Human Behaviour*, New York, NY: Macmillan.

Skocpol, T. (1985), 'Bringing the State Back In: Strategies of Analysis in Current Research', in Evans, P.B., Rueschemeyer, D. and Skocpol, T. (eds), *Bringing the State Back In*, Cambridge: Cambridge University Press.

Smith, G.W. (1981), 'Must Radicals Be Marxists? Lukes on Power, Contestability and Alienation', *British Journal of Political Science*, **11**, 405–25.

Smith, M. (1990a), *The Politics of Agricultural Support in Britain*, Aldershot: Dartmouth.

Smith, M. (1990b), 'From Policy Community to Issue Network: Salmonella in Eggs and the New Politics of Food', *Brunel Working Papers in Government*, **9**, Uxbridge: Brunel University.

Smith, M. (1990c), 'Pluralism, Reformed Pluralism and Neopluralism: The Role of Pressure Groups in Policy-Making', *Political Studies*, **38**, 302–22.

Sober, E. (1984), *The Nature of Selection: Evolutionary Theory in Philosophical Focus*, Cambridge, MA: Cambridge University Press.

Soloway, S.M. (1987), 'Elite Cohesion in Dahl's New Haven: Three Centuries of the Private School', in Domhoff and Dye (eds) (1987).

Stacy, M., Batstone, E., Bell, C. and Murcott, A. (1975), *Power, Persistence and Change: A Second Study of Banbury*, London: Routledge and Kegan Paul.

Steiner, H. (1975), 'Individual Liberty', *Proceedings of the Aristotelian Society*, LXXV (1974–75), 33–50.

Stigler, G. and Becker, G. (1977), 'De Gustibus Non Est Disputandum', *American Economic Review*, **67**, 76–90.

Stinchcombe, A. (1968), *Constructing Social Theories*, New York, NY: Harcourt, Brace and World.

Stone, C. (1976), *Economic Growth and Neighborhood Discontent*, Chapel Hill, NC: University of North Carolina Press.

Stone, C. (1980), 'Systemic Power in Community Decision Making: A Restatement of Stratification Theory', *American Political Science Review*, **74**, 978–90.

Stone, C. (1986), 'Power and Social Complexity', in Waste, R. (ed.) (1986).

Stone, C. (1987), 'Elite Distemper Versus the Promise of Democracy', in Domhoff and Dye (eds) (1987).

Stones, R. (1988), 'State-Finance Relations in Britain 1964–70: A Relational Approach to Contemporary History', *Essex Papers in Politics and Government*, Colchester: University of Essex.

Sutton, J. (1986), 'Non-Cooperation Bargaining Theory: An Introduction', *Review of Economic Studies*, **176**, 709–24.

Tannahill, R. (1975), *Food in History*, St Albans: Paladin.

Taylor, C. (1964), *The Explanation of Behaviour*, London: Routledge and Kegan Paul.

Taylor, C. (1985a), 'The Concept of a Person', in his *Human Agency and Language: Philosophical Papers 1*, Cambridge: Cambridge University Press.

Taylor, C. (1985b), 'What's Wrong with Negative Liberty', in his *Philosophy and the Human Sciences: Philosophical Papers 2*, Cambridge: Cambridge University Press.

Taylor, M. (1982), *Community, Anarchy and Liberty*, Cambridge: Cambridge University Press.

Taylor, M. (1987), *The Possibility of Cooperation*, Cambridge: Cambridge University Press.

Taylor, M. (1988), 'Rationality and Revolutionary Collective Action', in Taylor, M. (ed.), *Rationality and Revolution*, Cambridge: Cambridge University Press.

Therborn, G. (1982), 'What Does The Ruling Class Do When It Rules?', in Giddens, A. and Held, D. (eds), *Classes, Power and Conflict*, London: Macmillan.

Thomson, G. (1987), *Needs*, London: Routledge and Kegan Paul.

Tichenor, V. (2005), 'Maintaining Men's Dominance: Negotiating Identity and Power When She Earns More', *Sex Roles*, **53**, 3–4, 191–205.

Treier, S. and Jackman, S. (2008), 'Democracy as a Latent Variable', *American Journal of Political Science*, **52**, 1, 201–17.

Truman, D. (1951), *The Governmental Process*, New York, NY: Alfred A. Knopf.

Tversky, A. and Kahneman, D. (1981), 'The Framing of Decisions and the Rationality of Choice', *Science*, **211**, 453–8.

Urry, J. (1990), 'Lancaster: Small Firms, Tourism and the "Locality"', in Harloe, Pickvance and Urry (eds) (1990).

von Mises, L. (1949), *Human Action*, London: William Hodge.

Wade, R. (1987), *Village Politics: The Management of Common Property Resources in South India*, Cambridge: Cambridge University Press.

Waldner, D. (2012), 'Process Tracing and Causal Mechanisms', in Kincaid, H. (ed), *Oxford Handbook of the Philosophy of Social Science*, Oxford: Oxford University Press.

Walker, J. (1983), 'The Origins and Maintenance of Interest Groups in America', *American Political Science Review*, **77**, 390–406.

Wanniski, J. (1978), 'Taxes, Revenues and the "Laffer Curve"', *Public Interest*, **50**, 3–16.

Ward, H. (1979), 'A Behavioural Model of Bargaining', *British Journal of Political Science*, **9**, 201–18.

Ward, H. (1987), 'Structural Power – A Contradiction in Terms?', *Political Studies*, **35**, 593–610.

Waste, R. (ed.) (1986), *Community Power: Directions for Future Research*, Beverly Hills, CA: Sage.

Watson, J.B. (1930), *Behaviorism*, New York, NY: W. W. Norton.

Weber, M. (1978), *Economy and Society*, volumes 1 and 2, Roth, G. and Wittich, C. (eds), Berkeley, CA: University of California Press.

Wetzel, L. (2009), *Types and Tokens: On Abstract Objects*, Cambridge: Cambridge University Press.

White, A. (1975), *Modal Thinking*, Oxford: Basil Blackwell.

Whiteley, P. and Winyard, S. (1984), 'Influencing Social Policy: The Effectiveness of the Poverty Lobby in Britain', *Journal of Social Policy*, **12**, 1–26.

Williams, B. (1972), 'Deciding to Believe', in his *Problems of the Self*, Cambridge: Cambridge University Press.

Wilson, H. (1974), *The Labour Government 1964–70*, Harmondsworth: Penguin.

Winner, L. (1980), 'Do Artifacts Have Politics?', *Daedelus*, Winter, 121–36.

Wittgenstein, L. (1953), *Philosophical Investigations*, Oxford: Basil Blackwell.

Wolfinger, R. (1960), 'Reputation and Reality in the Study of Community Power', *American Sociological Review*, **25**, 636–44.

Wolfinger, R. (1971a), 'Nondecisions and the Study of Local Politics', *American Political Science Review*, **65**, 1063–80.

Wolfinger, R. (1971b), 'Rejoinder to Frey's "Comment"', *American Political Science Review*, **65**, 1102–4.

Wright, E.O. (1985), *Classes*, London: Verso.

Young, H. (1989), *One of Us*, London: Macmillan.

Index